# Structural
# Mechanics

**Macmillan College Work Outs for degree and professional students**

Dynamics
Electric Circuits
Electromagnetic Fields
Electronics
Elements of Banking
Engineering Materials
Engineering Thermodynamics
Fluid Mechanics
Heat and Thermodynamics
Mathematical Modelling Skills

Mathematics for Economists
Mechanics
Molecular Genetics
Numerical Analysis
Operational Research
Organic Chemistry
Physical Chemistry
Structural Mechanics
Waves and Optics

MACMILLAN WORK OUT SERIES

# Structural Mechanics

**Ray Hulse**
**Jack Cain**

MACMILLAN

First published 1991 by
MACMILLAN PRESS LTD
Houndmills, Basingstoke, Hampshire RG21 6XS
and London
Companies and representatives
throughout the world

ISBN 0–333–53549–9

A catalogue record for this book is available
from the British Library.

09  08  07  06  05  04  03
02  01  00  99  98  97  96  95


Printed in Malaysia

# Contents

# Preface

As a first-year undergraduate or HND student in civil or mechanical engineering or in building, you should be using this book as an aid to consolidating your understanding of the principles and application of structural mechanics.

This is not a standard textbook; it should be used as supporting material to your main course text and/or lecture material. Indeed, it has been written on the assumption that you have already studied, or are currently studying, structural mechanics, and that you now require additional help in applying the principles you have learnt to a range of problems typical of those that you will encounter in your end-of-year examinations. The greatest advantage can therefore be gained from this book by using it together with a main-stream textbook such as *Structural Mechanics* by J. A. Cain and R. Hulse (Macmillan Education, 1990).

Each chapter covers one main topic area that you may have studied on your course. At the beginning of each chapter the essential theory, important facts and relevant equations are presented and summarised in a '*Fact Sheet*'. This is the essential information that you must know and be familiar with. The main part of each chapter is devoted to a series of typical examination questions and worked solutions which will demonstrate the application of the theory to a wide range of typical examination standard problems. At the end of the chapter there are additional problems with answers given so that you can gain further practice in problem solving.

Structural mechanics, as with all mathematically based subjects, can only be completely mastered by attempting and solving problems for yourself. The more problems that you attempt the more confident you will become about the subject itself. Throughout the book you should therefore attempt to answer each problem yourself before looking at, and studying, the given solution. The presentation of the solutions is more detailed than would usually be expected under normal examination conditions. Additional explanation is given to most questions as an aid to understanding; the comments in **bold** type would not be expected to form part of the student's solution to a given problem. However, in an examination you should always attempt to set out the solution neatly, giving sufficient explanation to demonstrate to the examiner that the nature of the problem is fully understood and that the correct application of principles is being made.

In each chapter, immediately following the *Fact Sheet*, a list of the main symbols used throughout that chapter is given. These are generally commonly accepted standard symbols and reference to the list should minimise any difficulties if you have been using different symbols in your own study of structural mechanics. Commonly used units are quoted alongside each symbol although the actual units used in any given problem may differ depending on the nature of the problem and the information given. For example, units of stress may be quoted as N/mm$^2$ although in some cases stress may have been calculated in kN/m$^2$. In all problems, however, the over-riding consideration is that units must be consistent within any equation. It is important that all equations are dimensionally correct. The sign conventions used throughout the chapter are also quoted. Again, these have been chosen to be consistent with common practice.

In working through the examples and problems you should neither assume that all questions are equally demanding in terms of the time required to complete a solution nor equally valuable in terms of their contribution to the total marks in an examination. The format of the examination papers from which the questions have been selected varies somewhat and some questions are expected to take longer to do than others. In general, most of the questions should be capable of solution within 30–45 minutes. A few of the shorter questions are selected from papers where the solution is expected to be achieved within 15 minutes.

Finally, a comment about accuracy of calculations. The answers to many engineering problems can be slightly sensitive to the extent of numerical 'rounding off' used throughout the calculation. With the use of modern calculators it is possible to perform extensive complex calculations in a few calculator operations (or even a single operation). If intermediate calculations are performed and intermediate answers are 'rounded off' to, say, two or three decimal places before the result of the calculation is used in subsequent calculations, the final answer may vary slightly from the 'exact' answer. For this reason the reader should be prepared to recognise that slight differences between answers may only be an indication of 'rounding off' errors, whereas substantial differences will indicate some major source of error.

This book is intended to assist you in successfully passing your examinations and in forming the firm foundations of your further study. If you work your way through the majority of the given examples and problems we believe that both these objectives will be readily met. Good luck!

*Coventry, 1990*                                                                    R.H.
                                                                                     J.C.

# Acknowledgements

We are indebted to our academic colleagues from the universities listed below who assisted in the writing of this book by providing us with copies of recent first-year examination papers in structural mechanics and the syllabuses on which these questions were based.

In compiling this text we have endeavoured to select a range of questions from these papers which illustrate as many aspects of the subject as possible and which give a broad coverage of the topics which are most commonly studied in most undergraduate and HND engineering courses.

Except for minor alterations to maintain some uniformity of presentation, the questions are reproduced exactly in the form in which they were provided to us. In a few cases, particularly in earlier chapters, only parts of some questions have been reproduced in order to illustrate specific aspects of structural mechanics. Although we fully acknowledge the origins of the questions it must be stated that the solutions to these questions are entirely our own. While every effort has been made to ensure that the solutions given are accurate we, not the originators of the questions, accept the responsibility for any remaining inconsistencies or inaccuracies in the solutions which the reader may find.

Our thanks and acknowledgements go to:

Aberdeen University
Birmingham University
Cambridge University
Coventry University
Liverpool University
Manchester University
Nottingham Trent University

Nottingham University
Salford University
Sheffield University
University of Hertfordshire
University of Portsmouth
University of Westminster

# 1 Equilibrium of Rigid Structures

## 1.1 Contents

The principles of equilibrium ● The equations of equilibrium ● Support reactions for beams, plane frames, space frames, mass structures, arches and suspension cables.

The determination of the support reactions for a loaded structure is normally the first stage in any structural analysis and, hence, is normally the first part only of any examination question. Consequently, some of the following worked examples are only parts of actual examination questions; others have been specially designed to give practice in particular aspects. The reader will obtain further practice by working through the examples in later chapters.

## 1.2 The Fact Sheet

### (a) General Definition

A loaded structure is in *equilibrium* if it does not move as a rigid body. Rigid body movement can be either a translation (movement in a straight line) or a rotation or a combination of both.

In order for a structure to be in equilibrium, the effect of the loads, which tend to move or rotate the structure, must be balanced by reactive forces (*reactions*) developed at the *supports* upon which the structure is erected.

### (b) Plane Structures

A structure which lies within a single plane (i.e. in two dimensions) is a plane structure.

### (c)  Conditions for Equilibrium

The requirement that a plane structure does not move in *any* direction may be specified by stating that it must not move in any two perpendicular directions. Normally, but not essentially, the two directions are taken to be horizontal and vertical.

The structure will not move in any direction provided that there is no resultant force in that direction.

Thus, for no movement in the horizontal direction the algebraic sum of all the forces acting horizontally must be zero — that is,

$$\Sigma H = 0$$

Similarly, for no movement vertically

$$\Sigma V = 0$$

The requirement that a structure does not rotate in a plane implies that it can not rotate about any axis at right angles to that plane. Thus, there must be no resultant moment of force about any point in the plane. Hence, for no rotation in the plane the algebraic sum of the moment of all forces about any point in the plane must be zero — that is,

$$\Sigma M = 0$$

where the point in the plane about which moments are taken can be either within or external to the structure.

For complete equilibrium of the plane structure,

$\Sigma H = 0$:  the algebraic sum of all horizontal forces equals zero;
$\Sigma V = 0$:  the algebraic sum of all vertical forces equals zero;
$\Sigma M = 0$:  the algebraic sum of the moments of all forces about any point equals zero.

These are the three equations for the statical equilibrium of plane structures.

Sufficient supports and corresponding support reactions must be provided to enable the above-listed equations to be satisfied. Three unique equations are involved; thus, the magnitude of *three* unknown reactions (or fixing moments) may be determined. If a structure is provided with just sufficient supports (no more than three unknown reactions), it may be completely analysed by use of the above equations and is *externally statically determinate*. If there are too many support reactions, a solution is not possible by the use of the above equations alone and the structure is *statically indeterminate*. Such a structure is not dealt with in this book. If a structure has too few support reactions, it will move as a rigid body.


### (d)  Supports (Figure 1.1)

(i)  A *roller support* provides one reaction only, of unknown magnitude, and acting at right angles to the direction of motion of the rollers. Such a support permits linear movement in one direction only and also permits rotation.

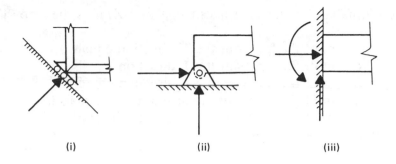

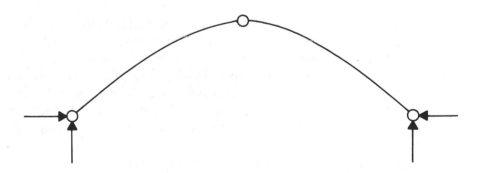 Note: The (i), (ii), (iii) labels belong to Figure 1.1 above.

(i)             (ii)             (iii)

**Figure 1.1**   Support types

(ii)   A *pinned support* provides a reaction of unknown magnitude and unknown direction which may be completely defined by determining the horizontal and the vertical components of reaction. Such a support prevents any linear movement but permits rotation.

(iii)   A *fixed support* provides a reaction of unknown magnitude and unknown direction and also a fixing moment (a total of three unknowns). Such a support prevents any linear movement and also prevents any rotation.

## (e)   Three-pinned Plane Structures

If a structure contains an internal pin (hinge) (e.g. as in Figure 1.2), such that one part of the structure can rotate about the pin independently of the rotation of the other part, then one additional unique equation of equilibrium may be written, since the sum of the moments of all forces on either part of the structure about that pin must be zero. This enables one additional unknown component of support reaction to be determined.

**Figure 1.2**   Three-pinned arch structure

## (f)   Space Structures

A three-dimensional structure is a *space structure*.

For a space structure, the sum of all the forces in each of three mutually perpendicular directions must be zero, and the sum of the moments of all forces about three mutually perpendicular axes ($X$, $Y$ and $Z$) must be zero. Thus,

$\Sigma X = 0$:　　the sum of the forces in the $X$ direction equals 0;
$\Sigma Y = 0$:　　the sum of the forces in the $Y$ direction equals 0;
$\Sigma Z = 0$:　　the sum of the forces in the $Z$ direction equals 0;
$\Sigma M_{XX} = 0$:　the sum of the moments about the $X$ axis equals 0;
$\Sigma M_{YY} = 0$:　the sum of the moments about the $Y$ axis equals 0;
$\Sigma M_{ZZ} = 0$:　the sum of the moments about the $Z$ axis equals 0.

## (g)　Mass Structures

A mass structure, such as the gravity dam shown in Figure 1.3, is one which

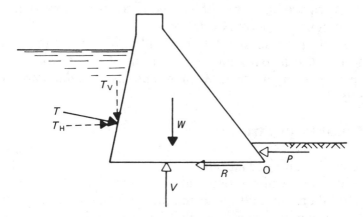

**Figure 1.3**　Gravity dam

depends upon its own weight to ensure equilibrium. Thus, for equilibrium,

$\Sigma V = 0$:　　the weight ($W$) of the structure and the vertical components $T_V$ of any loads ($T$) must be balanced by the vertically upward reaction ($V$) of the ground (or foundation) below the structure;

$\Sigma H = 0$:　　the tendency for the structure to slide horizontally under the action of any horizontal components ($T_H$) of load $T$ must be prevented by a reaction ($P$) from the ground behind the structure and/or by a frictional force ($R$) between the structure and the ground beneath it;

$\Sigma M_O = 0$:　The *overturning* moment of the loads about a probable centre of rotation (O) must be balanced by the *restoring moment* about the same point due to the self-weight of the structure.

A mass structure is normally designed so that its weight is greater than the minimum required for equilibrium in order to provide a factor of safety against overturning, where

$$\text{factor of safety} = \frac{\text{restoring moment}}{\text{overturning moment}}$$

Similarly, a factor of safety against sliding will be provided, where

$$\text{factor of safety} = \frac{\text{total reactive force resisting sliding}}{\text{total force tending to cause sliding}}$$

### (h) Principle of Superposition

If a structure is made of linear elastic material and is loaded by a combination of loads which do not strain the structure beyond the linear elastic range, then the resultant effect of the total load system on the structure is equivalent to the algebraic sum of the effect of each load acting separately.

## 1.3 Symbols, Units and Sign Conventions

$H$ = the horizontal reaction at a support (kN)
$M$ = moments taken about a support (kN m)
$V$ = the vertical reaction at a support (kN)

Clockwise moments are assumed to be positive.
Forces or components of forces acting to the right are assumed to be positive.
Forces or components of forces acting vertically upward are assumed to be positive.
The direction of action of known forces is indicated by solid arrowheads on the diagrams.
The assumed direction of action of unknown forces is indicated by open arrowheads on the diagrams.

## 1.4 Worked Examples

### Example 1.1

Beam ABCDE has a pinned support at A and a roller support at D. It carries three concentrated loads as shown in Figure 1.4. Determine the reactions.

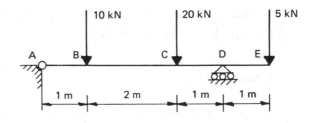

**Figure 1.4**  Load diagram

Support A can provide two components of reaction ($V_A$ and $H_A$). Support D, being a roller, provides only one reaction ($V_D$), which will act vertically (that is, at right angles to the direction of motion of the rollers). These reactions are shown on the free body diagram (Figure 1.5).

There are only three unknowns; thus, the structure is externally statically determinate and the unknowns can be determined.

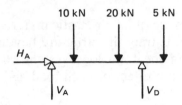

**Figure 1.5**  Free body diagram

### *Solution 1.1*

(1)  To determine $H_A$
      ($\Sigma H = 0$)   There are no horizontal loads; thus,

$$H_A = 0$$

(2)  To determine $V_D$
      Take moments about A:
      ($\Sigma M_A = 0$)

$$+(10 \times 1) + (20 \times 3) + (5 \times 5) - (V_D \times 4) = 0$$

$$\therefore V_D = +23.75 \text{ kN (i.e. 23.75 kN upwards)}$$

(3)  To determine $V_A$
      ($\Sigma V = 0$)

$$+V_A + V_D - 10 - 20 - 5 = 0$$
$$+V_A + (+23.75) - 35 \quad = 0$$

$$\therefore V_A = +11.25 \text{ kN}$$

(4)  Check by taking moments about D:

$$\Sigma M_D = +(V_A \times 4) - (10 \times 3) - (20 \times 1) + (5 \times 1)$$
$$= +(+11.25 \times 4) - 45 = 45 - 45 = 0$$

$$\therefore \text{ correct}$$

This last calculation provides a useful check on the mathematical accuracy of the previous two calculations.

### Example 1.2

Beam ABCD has a pinned support at A and a roller support at C. It carries two concentrated loads of 15 kN each and a uniformly distributed load of 2 kN/m over the right-hand half, as shown in Figure 1.6. Determine the reactions.

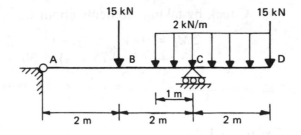

**Figure 1.6**

A free body diagram has been drawn (Figure 1.7) to show the reactions which are to be determined.

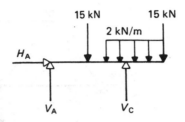

**Figure 1.7** Free body diagram

## *Solution 1.2*

(1)  To determine $H_A$

($\Sigma H = 0$)   There are no horizontal loads.

$$\therefore \underline{H_A = 0}$$

(2)  To determine $V_C$

Take moments about A:

Note that the moment of the U.D. load is the resultant total U.D. load ($2 \times 3 = 6$ kN) multiplied by the distance from A to the line of action of that resultant (i.e. 4.5 m)

($\Sigma M_A = 0$)

$$+(15 \times 2) - (V_C \times 4) + (2 \times 3 \times 4.5) + (15 \times 6) = 0$$

$$\therefore \underline{V_C = +36.75 \text{ kN}}$$

(3)  To determine $V_A$

($\Sigma V = 0$)

$$+V_A - 15 + V_C - (2 \times 3) - 15 = 0$$
$$+V_A - 15 + (+36.75) - 6 - 15 = 0$$

$$\therefore \underline{V_A = -0.75 \text{ kN}} \text{ (i.e. 0.75 kN downwards)}$$

(4) Check by taking moments about C:

$$\Sigma M_C = +(V_A \times 4) - (15 \times 2) + (2 \times 3 \times 0.5) + (15 \times 2)$$
$$= +(-0.75 \times 4) - 30 + 3 + 30 = 0$$

$$\therefore \text{ correct}$$

## Example 1.3

The truss shown in Figure 1.8 has a pinned support at A and a roller support at B. Determine the reactions at the supports when loads act on the truss as shown.

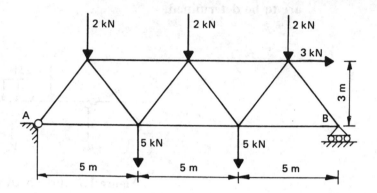

**Figure 1.8**

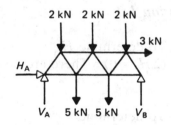

**Figure 1.9**   Free body diagram

## Solution 1.3

(1) To determine $H_A$
   $(\Sigma H = 0)$

$$+H_A + 3 = 0$$

$$\therefore H_A = -3.00 \text{ kN}$$

(2) To determine $V_B$
   Take moments about A:
   $(\Sigma M_A = 0)$

$$+(2 \times 2.5) + (5 \times 5) + (2 \times 7.5) + (5 \times 10) + (2 \times 12.5) + (3 \times 3)$$
$$- (V_B \times 15) = 0$$

$$\therefore V_B = +8.60 \text{ kN}$$

8

(3) To determine $V_A$
$$(\Sigma V = 0)$$
$$+V_A - 2 - 5 - 2 - 5 - 2 + V_B = 0$$
$$V_A - 16 + (+8.60) = 0$$
$$\therefore V_A = +7.40 \text{ kN}$$

(4) Check by taking moments about B:
$$\Sigma M_B = +(V_A \times 15) - (2 \times 12.5) - (5 \times 10) - (2 \times 7.5) - (5 \times 5)$$
$$- (2 \times 2.5) + (3 \times 3) = +(+7.4 \times 15) - 111 = 0$$
$$\therefore \text{correct}$$

**An alternative solution might be to make use of the principle of superposition to determine the vertical reactions, as follows:**

(1) Vertical loading acting alone

Since the vertical loading is symmetrically positioned on the truss, then
$$V_A = V_B = (\text{total vertical load})/2$$
$$= (2 + 5 + 2 + 5 + 2)/2 \quad = +16/2$$
$$\therefore V_A = V_B = +8.00 \text{ kN}$$

(2) Horizontal load acting alone
Taking moments about A:
$$(\Sigma M_A = 0)$$
$$+(3 \times 3) - (V_B \times 15) = 0$$
$$\therefore V_B = +0.60 \text{ kN}$$

$$(\Sigma V = 0)$$
$$V_A + V_B = 0, \qquad V_A + (+0.60) = 0$$
$$\therefore V_A = -0.60 \text{ kN}$$

(3) Superimposing the effect of the two loadings:
total $V_A = +8.00 - 0.60$
$$\therefore V_A = +7.40 \text{ kN}$$

and
total $V_B = +8.00 + 0.60$
$$\therefore V_B = +8.60 \text{ kN}$$

## Example 1.4

Find the reaction components at the supports for the truss shown in Figure 1.10.

(Sheffield University)

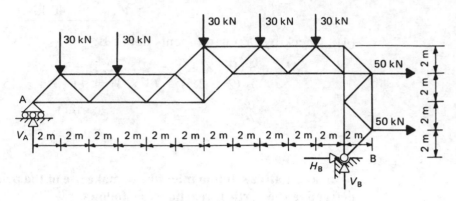

**Figure 1.10**

This problem looks very complex, but the procedure for solution is as in the previous example. Note that the supports are at different levels; thus, the horizontal component at support B has a moment about the other support (A). $\Sigma M_A$ will thus provide an equation with two unknowns. It is consequently better to take moments about support B. The drafting of the moment equations may be facilitated if the assumed directions of the support reactions are sketched on the diagram before proceeding. The open-headed arrows at A and B have accordingly been added to the original diagram to show the presumed reactions.

### *Solution 1.4*

(1) To determine $V_A$
Taking moments about B:
$(\Sigma M_B = 0)$
$+(V_A \times 22) - (30 \times 20) - (30 \times 16) - (30 \times 10) - (30 \times 6)$
$- (30 \times 2) + (50 \times 6) + (50 \times 2) = 0$
$$+(V_A \times 22) + 400 - 1620 = 0$$
$$\therefore \underline{V_A = +55.455 \text{ kN}}$$

(2) To determine $V_B$
$(\Sigma V = 0)$
$$+V_A + V_B - (5 \times 30) = 0$$
$$+(+55.455) + V_B - 150 = 0$$
$$\therefore \underline{V_B = +94.545 \text{ kN}}$$

(3) To determine $H_B$
$(\Sigma H = 0)$
$$+H_B + 50 + 50 = 0$$
$$\therefore \underline{H_B = -100.00 \text{ kN}}$$

(4)   Check by taking moments about A:

$$\Sigma M_A = +(30 \times 2) + (30 \times 6) + (30 \times 12) + (30 \times 16) + (30 \times 20)$$
$$+ (50 \times 2) - (50 \times 2) - (H_B \times 4) - (V_B \times 22)$$
$$= +1680 - (-100.00 \times 4) - (+94.545 \times 22) = 0$$

$$\therefore \text{correct}$$

**Note that in this example the values of the reactions have been quoted to three places of decimals. If the value of $V_B$ had been rounded up to two places of decimals, then the check calculation would not have proved to be correct. In general, however, it is quite sufficient to quote answers to two places of decimals.**

### Example 1.5

Figure 1.11 shows a pin-jointed structure, which has pinned supports at both A and C, supporting a vertical and a horizontal load.

(a)   It has four reactive components, two at each support A and C; nevertheless, it is statically determinate both internally and externally. Explain why this is so.

(b)   Calculate in terms of $W$ and $L$ the vertical and horizontal reactive components for the loading shown.

(Nottingham University)

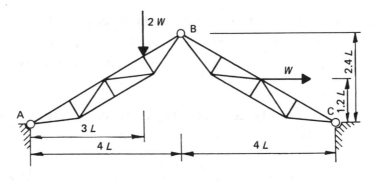

**Figure 1.11**

### Solution 1.5

(a)

**The answer to the question of external determinacy follows from the Fact Sheet at the beginning of this chapter. The answer in respect of internal determinacy is related to the Fact Sheet for Chapter 2 and the reader is advised to refer to that Fact Sheet if in any doubt about the following answer.**

For a structure to be externally statically determinate, it must normally have no more than a total of three unknown components of reaction or moments, since only three unique equations for equilibrium can be written. In this case, however, there is a central pin and a fourth equation may be written, since the sum of the moments of all the forces on either the left- or right-hand part of the structure about that pin must be zero for equilibrium. Thus, four unknown reactive components may be determined and the structure is externally statically determinate.

For a frame to be internally statically determinate, the number of members, $M$, and the number of joints, $J$, in that frame must be related by the expression $M = 2J - 3$. In this case each part of the structure is a frame in which $M = 13$ and $J = 8$. These numbers satisfy the requirement that $M = 2J - 3$. In addition, the members must be correctly positioned within the frame (the structure should be completely triangulated). Inspection shows this to be so; thus, both parts of the structure are, and, hence, the complete structure is, internally statically determinate.

(b)

(1)  To determine $V_C$, consider the whole frame and take moments about A:
$(\Sigma M_A = 0)$

$$+(2W \times 3L) + (W \times 1.2L) - (V_C \times 8L) = 0$$

$$\therefore V_C = +0.90\ W$$

(2)  To determine $V_A$, consider the whole frame and resolve vertically:
$(\Sigma V = 0)$

$$+V_A + V_C - 2W = 0$$
$$+V_A + (+0.90W) - 2W = 0$$

$$\therefore V_A = +1.10W$$

(3)  Check by taking moments about C:

$$\Sigma M_C = +(V_A \times 8L) + (W \times 1.2L) - (2W \times 5L)$$
$$= + (+1.1W \times 8L) + (W \times 1.2L) - (2W \times 5L) = 0$$

$$\therefore \text{correct}$$

(4)  To determine $H_A$, consider the left-hand part only and take moments about B:

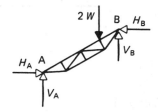

**Figure 1.12**  Left-hand part

$(\Sigma M_B = 0)$

$$+(V_A \times 4L) - (2W \times L) - (H_A \times 2.4L) = 0$$
$$+(+1.1W \times 4L) - (2W \times L) - (H_A \times 2.4L) = 0$$

$$\therefore H_A = +1.00W$$

(5) To determine $H_C$, consider the whole frame and resolve horizontally:
$$(\Sigma H = 0)$$
$$+H_A + H_C + W = 0$$
$$+(+1.0W) + H_C + W = 0$$
$$\therefore \underline{\underline{H_C = -2.00W}}$$

## Example 1.6

Figure 1.13 shows a three-pinned arch which is formed from two circular segments and carries two point loads. Construct free body diagrams for (i) part AB (ii) part BC and (iii) part AX, giving the forces at X in terms of shear and thrust.

(University of Portsmouth)

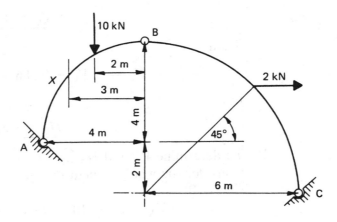

**Figure 1.13**

## Solution 1.6

**(i), (ii) To be complete, the free body diagrams should show the magnitude and direction of the reactive components. Consequently, the first stage of the solution must be to determine the reactions as in the previous example.**

**Note that, in this case, it is not possible to calculate the value of $V_C$ directly by taking moments about A, since the two supports are at different levels.**

(1)  To determine $V_C$ and $H_C$
Consider the right-hand part (BC) only (see Figure 1.14) and take moments about B:
$$(\Sigma M_B = 0)$$
$$-\{2 \times (6 - 6 \sin 45°)\} - (H_C \times 6) - (V_C \times 6) = 0$$
$$\therefore \underline{\underline{H_C = -V_C - 0.586}}$$

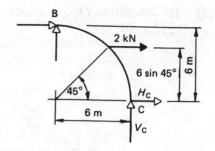

**Figure 1.14** Right-hand part

Consider the whole structure and take moments about A:

$$(\Sigma M_A = 0)$$
$$+(10 \times 2) + \{2 \times (6\sin45° - 2)\} - (H_C \times 2) - (V_C \times 10) = 0$$
$$+20 + 4.485 - 2 \times (-V_C - 0.586) - (V_C \times 10) = 0$$
$$\therefore V_C = +3.207 \text{ kN}$$

Then

$$H_C = -V_C - 0.586 = -(+3.207) - 0.586$$

i.e.

$$H_C = -3.793 \text{ kN}$$

(2)  To determine $V_A$ and $H_A$
Consider the whole structure:
$$(\Sigma V = 0)$$
$$+V_A + V_C - 10 = 0 + V_A + (+3.207) - 10 = 0$$
$$\therefore V_A = +6.793 \text{ kN}$$

$$(\Sigma H = 0)$$
$$+H_A + H_C + 2 = 0. \qquad +H_A + (-3.793) + 2 = 0$$
$$\therefore H_A = +1.793 \text{ kN}$$

(3)  Check by taking moments about C for the whole structure:

$$\Sigma M_C = +(H_A \times 2) + (V_A \times 10) - (10 \times 8) + (2 \times 6\sin45°)$$
$$= +(+1.793 \times 2) + (+6.793 \times 10) - 80 + 8.485 = 0$$

$$\therefore \text{ correct}$$

**The free body diagrams for part AB and part BC are shown in Figures 1.15 and 1.16, respectively. Note that the reactive forces acting at B on part AB are**

$$V_B = 10.000 - 6.793 = 3.207 \text{ kN}$$

**and**

$$H_B = -H_A = -1.793 \text{ kN}$$

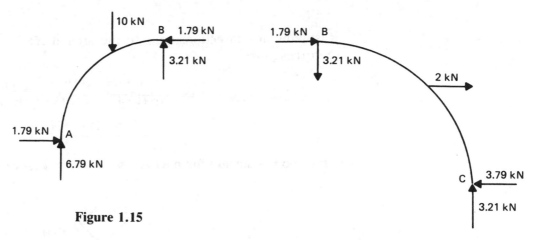

**Figure 1.15**

**Figure 1.16**

**The reactive forces acting at B on part BC are equal in magnitude but opposite in direction to these.**

(iii)
(1)  To determine thrust and shear at $X$
Refer to Figure 1.17:

**The part A$X$ of the structure is in equilibrium under the action of $V_A$ and $H_A$ acting at support A together with a thrust $T$ (a direct force) acting along the longitudinal axis of the arch (i.e. tangential to the arch) at $X$, and a shearing force $S$ acting at right angles to the axis of the arch (i.e. radially to the arch) at $X$. The direction of action of the forces $S$ and $T$, as indicated in Figure 1.17, is taken as positive in the following calculations.**

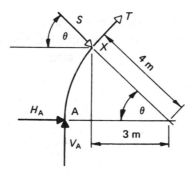

**Figure 1.17**

Resolving in a direction tangential to the arch at $X$.
($\Sigma$ forces $= 0$)

$$T + H_A \sin\theta + V_A \cos\theta = 0$$

where, from the geometry of the diagram, $\theta = \cos^{-1}(3/4) = 41.41°$. Thus,

$$T + 1.793 \sin 41.41° + 6.793 \cos 41.41° = 0$$

$$\therefore \underline{T = -6.281 \text{ kN}}$$

15

Resolving in a direction radial to the arch at $X$:
($\Sigma$ forces = 0)

$$S + H_A \cos\theta - V_A \sin\theta = 0$$
$$S + 1.793 \cos41.41° - 6.793 \sin41.41° = 0$$

$$\therefore \underline{S = 3.148 \text{ kN}}$$

**The free body diagram for part A$X$ is shown in Figure 1.18.**

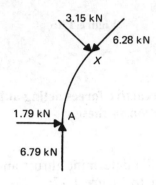

**Figure 1.18**

## Example 1.7

A concrete dam has a trapezoidal cross-section as shown in Figure 1.19. Water is impounded to a depth of 10 m.

(a)  If the density of water is 1000 kg/m³, determine:
 (i)  the horizontal thrust on the dam expressed in newtons per metre length of dam;

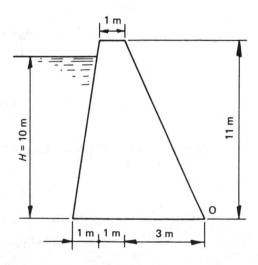

**Figure 1.19**

(ii)   the total thrust on the dam per metre length;
(iii)  the position of the centre of pressure on the wetted face of the dam;
(iv)   the overturning moment about the toe (O).

(b)  If the density of the concrete is 2400 kg/m³, determine:

(i)   the position of the centre of gravity of the dam;
(ii)  the factor of safety against overturning.

(Coventry University)

This is a mass structure problem in which the structure may fail by overturning about the toe O, or by sliding horizontally. In this case the question does not consider the possibility of failure by sliding. The wording of the question guides the candidate through the sequential steps of the solution.

*Solution 1.7*

(a)
(i)   The horizontal thrust per metre length is given by

$$(T_H) = \rho g H^2/2$$
$$= (1000 \times 9.81 \times 10^2)/2$$
$$= 490.50 \text{ kN}$$

The hydrostatic pressure increases linearly from zero at the surface of the water to a maximum value of $\rho g H$ at the base. The average pressure is $\rho g H/2$ and the total horizontal thrust (average pressure × the area of the face of the dam projected onto the vertical plane) is $(\rho g H/2) \times (H \times 1) = \rho g H^2/2$ per metre length of dam.

(ii)   The total thrust $(T)$ per metre length is given by

$$T = \text{average pressure} \times \text{wetted area of face of dam}$$

The wetted area per metre length of the dam equals the length of the sloping face of the dam which is in contact with the water (i.e. length DE) multiplied by one metre (see Figure 1.20).

$$DE = (10^2 + 0.909^2)^{1/2} = 10.0412$$

Thus

$$T = (\rho g H/2) \times (10.0412 \times 1)$$
$$= (1000 \times 9.81 \times 10/2) \times 10.0412 \times 10^{-3}$$
$$= 492.52 \text{ kN}$$

An alternative method of determining the total thrust is to calculate the vertical component of thrust, $T_V$, and compound it with the horizontal thrust, $T_H$, as follows:

Vertical component of thrust = weight of the wedge DEF

$$T_V = \text{volume of wedge} \times \text{unit weight of water}$$
$$= \tfrac{1}{2} \times (10 \times 0.909 \times 1) \times 1000 \times 9.81)$$
$$\times (1000 \times 9.81) \times 10^{-3}$$
$$= 44.59 \text{ kN}$$

$$\text{Total thrust} = (T_H^2 + T_V^2)^{1/2}$$
$$= (490.50^2 + 44.59^2)^{1/2}$$
$$= 492.52 \text{ kN}$$

(iii)  The centre of pressure is at 2/3 total depth of water

$$= 2/3 \times 10$$
$$= 6.67 \text{ m vertically below the water surface}$$

**The hydrostatic pressure acting on face DE increases linearly from zero at the surface of the water (E) to $\rho g H$ at the base (D). Thus, the load distribution diagram for the face DE is triangular and the resultant thrust which acts through the centroid of the load distribution diagram will act at two-thirds of the depth.**

(iv)  Taking moments of all overturning forces about the toe O:

overturning moment
$$= +(T_H \times 3.33) - (T_V \times 4.70) \qquad \textbf{(see Figure 1.20)}$$
$$= +(490.50 \times 3.33) - (44.59 \times 4.70)$$
$$= + 1423.79 \text{ kNm}$$

(b)

(i)  To determine the position of the centre of gravity of the dam with respect to the toe (O), take moments of the area of parts A, B and C about O (see Figure 1.20).

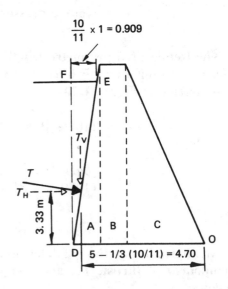

**Figure 1.20**

| Part | Area ($A$) | Distance ($x$) from O to centroid | $Ax$ |
|------|------------|-----------------------------------|------|
| A | $\frac{1}{2}(1 \times 11) = 5.50$ | 4.333 | 23.83 |
| B | $(1 \times 11) = 11.00$ | 3.500 | 38.50 |
| C | $\frac{1}{2}(3 \times 11) = 16.50$ | 2.000 | 33.00 |
| | $\Sigma A \quad = 33.00$ | | $\Sigma Ax = 95.33$ |

The centre of gravity is to the left of O by distance $\bar{x}$, where

$$\bar{x} = \frac{\Sigma Ax}{\Sigma A}$$

$$= \frac{95.33}{33.00} = 2.889 \text{ m}$$

(ii) The restoring moment per metre length tending to prevent the dam rotating about O

$\qquad$ = weight of dam × horizontal distance from (O)
$\qquad$ = volume × unit weight × $\bar{x}$
$\qquad$ = $(33.00 \times 1) \times (2400 \times 9.81) \times 2.889 \times 10^{-3}$
$\qquad$ = 2244.61 kNm

The factor of safety

$$= \frac{\text{restoring moment}}{\text{overturning moment}}$$

$$= \frac{2244.61}{1423.79} = 1.58$$

## Example 1.8

The arrangement shown in Figure 1.21 is used to withdraw the block B, which is between the fixed surface C and the block A. The self-weight of each part is

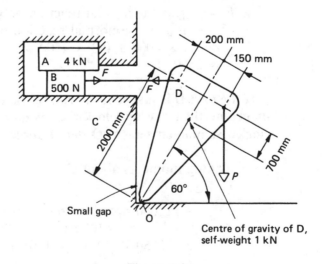

**Figure 1.21**

shown on the diagram and the coefficients of friction are $\mu_{BC} = 0.3$, $\mu_{DC} = 0.5$ and $\mu_{AB} = 0.3$.

(i) Determine whether or not the base of part D will slip (and then rest against the vertical surface of C) before block B is moved.

(ii) Assuming that the geometry remains essentially the same, whether or not slipping occurs between D and C, draw a free body diagram for part D at the instant that part B begins to move.

(University of Portsmouth)

**This arrangement is a mechanism, but the problem is concerned with a situation when the parts are only on the point of moving and not actually moving. In such a situation, when the parts are static, the arrangement is effectively a structure and the analysis of forces and reactions is precisely the same as for a structure. The solution involves a clear appreciation of basic principles relating to forces in equilibrium.**

*Solution 1.8*

(i) If $F$ is the tensile force in the horizontal link between B and D, then, when B is just about to move, $F$ equals the limiting frictional force between A and B plus the limiting frictional force between B and C (see Figure 1.22).

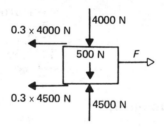

**Figure 1.22**

That is,

$F = \mu_{AB} \times$ the normal reaction between A and B
$\quad + \mu_{BC} \times$ the normal reaction between B and C
$\quad = 0.3 \times 4000 + 0.3 \times 4500$
$\quad = 2550$ N

**If the part D is just about to rotate about the contact point (O) indicated on the diagram, then the anticlockwise moment of $F$ about O will be equal to the clockwise moments about O due to force $P$ and to the self-weight of part D.**

Taking moments about O:
$(\Sigma M_O = 0)$

$$-F(2000 \sin 60° + 200 \sin 30°) + P(2000 \cos 60° + 150 \cos 30°)$$
$$+1000\{(2000 - 700)\cos 60°\} = 0$$
$$\therefore -2550(1832) + P(1130) + 1000(650) = 0$$

$$\therefore P = 3559 \text{ N}$$

**For vertical equilibrium of part D the vertical component of reaction at O will equal the weight of D plus force *P*; thus,**

vertical reaction at O = 1000 + 3559 = 4559 N

The maximum frictional force which can be developed at O (i.e. the maximum possible horizontal component of the reaction at O)

$$= \mu_{DC} \times \text{vertical reaction at O}$$
$$= 0.5 \times 4559$$
$$= 2279.5 \text{ N}$$

The horizontal component of reaction required at O to maintain equilibrium = $F$ = 2550 N. This is greater than the maximum frictional force which can be developed; thus, block D will slip to the left.

(ii)
**After block D has slipped and made contact with the vertical face of C, a horizontal reaction will be developed at the new point of contact to make up the deficiency in horizontal restraint. The value of this reaction is given by**

reaction at vertical face of C = 2550 − 2279.5 = 270.5 N

The free body diagram for part D at the instant that part B begins to move is drawn in Figure 1.23.

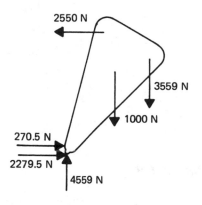

2550 N

3559 N

1000 N

270.5 N

2279.5 N

4559 N

**Figure 1.23**

## Example 1.9

Figure 1.24 shows a non-uniform cable hanging under gravity between two supports at the same level. The right-hand part, of projected length $L$, is a heavy flexible chain of weight $w$ per unit length, while the left-hand part, also of projected length $L$, is a light wire of negligible weight. The central dip ($d$) is small in comparison with the span. The vertical and horizontal scales of the sketch are not equal.

(a)  Use the conditions of statical equilibrium to determine the vertical components of the end reactions; and hence, by considering the equilibrium of a

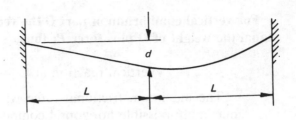

**Figure 1.24**

suitable piece, determine the horizontal component of the cable tension and thus an approximate expression for the tension ($T$) throughout the cable.

(b) By considering the equilibrium of another suitable piece find the location of the lowest point and the dip of the cable at that point.

(Cambridge University)

**The wording of the question makes it clear that the horizontal projected lengths of the cable may be taken to be the same as the actual length of cable.**

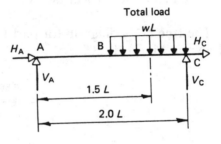

**Figure 1.25**

*Solution 1.9*

(a)

(1)  Considering the whole cable
Weight of cable = $w$ per unit length over right-hand half (BC)
Then, by taking moments about A,
($\Sigma M_A = 0$)

$$+(wL \times 1.5L) - (V_C \times 2L) = 0$$

$$\therefore \underline{V_C = +0.75wL}$$

Resolving vertically:
($\Sigma V = 0$)

$$+V_A + V_C - wL = 0$$
$$+V_A + 0.75wL - wL = 0$$

$$\therefore \underline{V_A = +0.25wL}$$

(2)    Considering the right-hand half (see Figure 1.26)

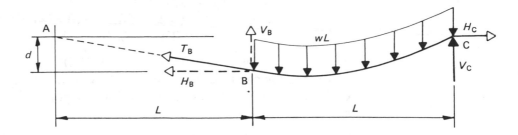

**Figure 1.26**

**The right-hand part is held in equilibrium by reactive components $V_C$ and $H_C$ at the support C together with a tension $T_B$ in the cable at B. $T_B$ may be determined by calculating its components $H_B$ and $V_B$, as indicated on Figure 1.26. Note that since the left-hand half is weightless, it will adopt the configuration of a straight line of slope $d/L$, as indicated by the dotted line in Figure 1.26. Since it is known that the cable will be in tension, the direction of $T_B$ is known; hence, $H_B$ has been drawn acting to the left.**

Resolving vertically:
$(\Sigma V = 0)$

$$+V_B + V_C - wL = 0$$
$$V_B + 0.75wL - wL = 0$$
$$\therefore V_B = +0.25wL$$

Taking moments about C:
$(\Sigma M_C = 0)$

$$+(V_B \times L) + (H_B \times d) - (wL \times 0.5L) = 0$$
$$+0.25wL^2 + H_B \times d - 0.5wL^2 = 0$$
$$\therefore H_B = \frac{+0.25wL^2}{d}$$

But the horizontal component of the tension is constant throughout, since there is no horizontal component of loading acting anywhere along the cable. Hence,

horizontal component of cable tension

$$H = +\frac{0.25wL^2}{d}$$

**This result could also be obtained by considering the geometry at B. The direction of $T_B$ is known to be at a slope of $d/L$; thus, $V_B/H_B$ must equal $d/L$. $V_B$ has been determined; thus, $H_B$ can be calculated.**

(3)  The cable tension in the left-hand half (AB) will be constant and of value equal to $T_B$, where

$$T_B = (H_B^2 + V_B^2)^{1/2}$$
$$= \{(0.25wL^2/d)^2 + (0.25wL)^2\}^{1/2}$$
$$= 0.25wL\{(L/d)^2 + 1\}^{1/2}$$

The cable tension in the right-hand half (BC) will vary along the length and an expression can be derived to give the tension at any point between B and C. Considering Figure 1.27, the tension at a point $X$, distance $x$ from C, will be

$$T_X = (H_X^2 + V_X^2)^{1/2}$$
$$= \underline{\{(0.25wL^2/d)^2 + (wx - 0.75wL)^2\}^{1/2}}$$

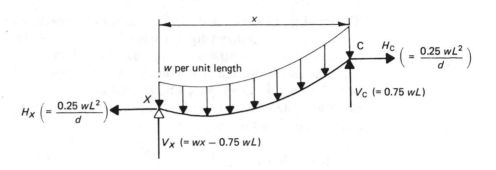

**Figure 1.27**

(b)

(1)  To locate the lowest point

Consider Figure 1.28, which shows the part of the cable to the right of the lowest point (P). The dimension $D$ shown in the figure is the maximum dip.

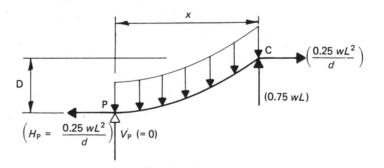

**Figure 1.28**

**Note that at the lowest point the cable will be horizontal, and the tension ($T_P$) will be horizontal and will have zero vertical component. Consequently, the component $V_P$ is zero.**

24

Resolving vertically:
($\Sigma V = 0$)

$$+V_P + V_C - \text{weight of cable} = 0$$
$$0 + 0.75wL - wx = 0$$
$$\therefore \underline{x = 0.75L}$$

and the weight of the cable CP = $(0.75L) \times w = 0.75wL$.
Taking moments about C:
($\Sigma M_C = 0$)

$$+(H_P \times D) - (0.75wL \times 0.375L) = 0$$
$$\therefore \frac{0.25wL^2}{d} \times D - 0.281wL^2 = 0$$

$$\therefore D = \frac{0.281d}{0.25} = 1.125d$$

$\therefore$ Maximum dip is given by $\underline{D = 1.125d}$

**Example 1.10**

The space frame shown in Figure 1.29 has a ball-jointed support at A. The support at B is free to move in any direction on the horizontal plane and the support at C is free to move in direction AC only. Calculate the value of the reactions at A, B and C.

[*Hint*  Take A as the origin of the three axes; the $X$ axis in direction AC, the $Z$ axis at right angles to the $X$ axis in the horizontal plane and the $Y$ axis vertical.]

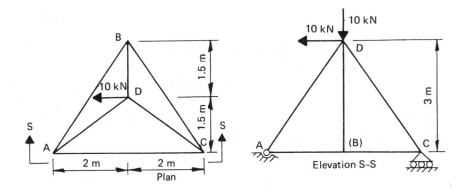

**Figure 1.29**

*Solution 1.10*

**The first step in the solution is to identify the reactions whose values are to be determined.**

Figure 1.30 shows the six components of reaction which are to be determined.

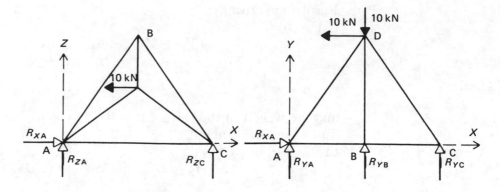

**Figure 1.30**

The ball joint at A provides reactive components in three directions ($X$, $Y$ and $Z$). Support B is restrained only in the $Y$ direction and, hence, provides only one reactive component (in the $Y$ direction). Support C is restrained in two directions ($Y$ and $Z$) and, hence, provides two reactive components (in the $Y$ and $Z$ directions).

Taking moments about the $X$ axis:
($\Sigma M_{XX} = 0$)
$$-(10 \times 1.5) + (R_{YB} \times 3.0) = 0$$
$$\therefore \underline{R_{YB} = +5.00 \text{ kN}}$$

Taking moments about the $Z$ axis:
($\Sigma M_{ZZ} = 0$)
$$+(10 \times 2) - (10 \times 3) - (R_{YB} \times 2) - (R_{YC} \times 4) = 0$$
$$+20 - 30 - (+5 \times 2) - (R_{YC} \times 4) = 0$$
$$\therefore \underline{R_{YC} = -5.00 \text{ kN}}$$

Taking moments about the $Y$ axis:
($\Sigma M_{YY} = 0$)
$$+(10 \times 1.5) + (R_{ZC} \times 4) = 0$$
$$\therefore \underline{R_{ZC} = -3.75 \text{ kN}}$$

Resolving in the $X$ direction:
($\Sigma X = 0$)
$$+R_{XA} - 10 = 0$$
$$\therefore \underline{R_{XA} = +10.00 \text{ kN}}$$

Resolving in the $Z$ direction:
($\Sigma Z = 0$)
$$+R_{ZA} + R_{ZC} = 0 \qquad +R_{ZA} + (-3.75) = 0$$
$$\therefore \underline{R_{ZA} = +3.75 \text{ kN}}$$

Resolving in the $Y$ direction (**vertical**):
($\Sigma Y = 0$)

$$+R_{YA} + R_{YB} + R_{YC} - 10 = 0$$
$$+R_{YA} + (+5.00) + (-5.00) - 10 = 0$$
$$\therefore \underline{R_{YA} = +10.00 \text{ kN}}$$

**Moments have been taken as positive if they act in a clockwise direction about an axis when viewed from A in the positive direction of the axis.**

## 1.5 Problems

**1.1** The plane pin-jointed truss shown in Figure P1.1 carries a horizontal load at D of $2W$ and a vertical load at G of $W$. Evaluate all the support reactions.

(University of Portsmouth)

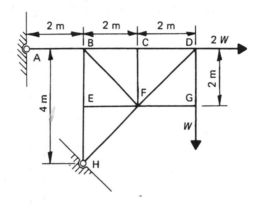

**Figure P1.1**

**1.2** Figure P1.2 shows a beam AC with a pinned support at A and on roller supports at B and C. A pin at P connects the two parts of the beam together. Determine the values of the reactions at A, B and C.

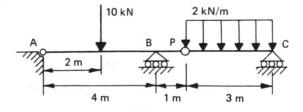

**Figure P1.2**

**1.3** Determine the magnitudes and directions of the horizontal and vertical components of reaction at A and F in the pin-jointed frame shown in Figure P1.3.

(Nottingham Trent University)

27

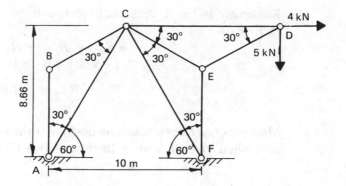

**Figure P1.3**

**1.4** Figure P1.4 shows a concrete retaining wall supporting soil to a depth of 2 m. The soil may be considered as exerting a total horizontal thrust on the face of the wall of 5.1 kN per metre length of wall. If the density of concrete is 2400 kg/m$^3$ and if the coefficient of friction between the base of the wall and the soil is 0.35, determine the factors of safety against overturning and sliding.

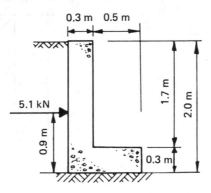

**Figure P1.4**

**1.5** Figure P1.5 shows a cantilevered pin-jointed frame supporting a vertical load at

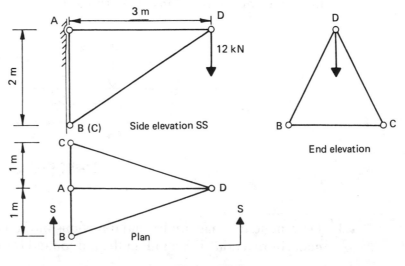

**Figure P1.5**

28

D. Support A is pinned (not capable of linear movement in any direction), B is free to move vertically only and C is free to move in any direction in the vertical plane. Determine the values and directions of all the support reactions.

[*Hint*   Take B as the origin of three axes; the $X$ axis in direction BC, the $Z$ axis at right angles to the $X$ axis and in the horizontal plane, and the $Y$ axis vertical.]

## 1.6   Answers to Problems

**1.1**   $H_A = -3W$   (3$W$ to the left); $H_H = +W$   ($W$ to the right);
$V_H = +W$   ($W$ vertically upwards)

**1.2**   $V_A = +4.25$ kN; $V_B = +8.75$ kN; $V_C = +3.00$ kN

**1.3**   $H_A = -3.44$ kN; $V_A = -5.96$ kN
$H_F = -0.56$ kN; $V_F = +10.96$ kN

**1.4**   Factor of safety against overturning is 2.19.
Factor of safety against sliding is 1.21.

**1.5**   $R_{ZA} = -18.0$ kN; $R_{XA} = 0.0$ kN; $R_{YA} = +12.0$ kN
$R_{ZB} = +9.0$ kN; $R_{XB} = 0.0$ kN
$R_{ZC} = +9.0$ kN         (see Figure P1.6)

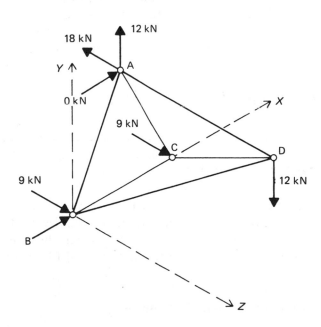

**Figure P1.6**

# 2 Pin-jointed Frame Structures

## 2.1 Contents

Determination of the magnitude and sense of the forces in the members of pin-jointed frames • Method of resolution at joints • Method of sections • Three-pinned frames • Space frames • Tension coefficients • Graphical method.

The questions in this chapter are concerned with the determination of the forces in the members of pin-jointed frames. Such analysis is required at the early stages in the design of such structures. Thus, you will find similar methods of analysis used in the questions in Chapter 12. All the examples involve frames which are internally and externally statically determinate (see Chapter 1).

## 2.2 The Fact Sheet

### (a) General Definition

A pin-jointed frame is a structure constructed from a number of straight members connected together at their ends by frictionless pinned joints. It is assumed that all external loads are applied at the joints of the structure such that the internal members are subjected to axial forces only.

### (b) Internal Statical Determinacy

A plane frame is internally statically determinate (that is, capable of solution using the equations of equilibrium) if it satisfies the relationship

$$M = 2J - 3$$

where $M$ = number of members and $J$ = number of joints. Similarly, for a space frame in equilibrium, $M = 3J - 6$.

### (c) Positioning of Members

Not only must the frame satisfy the relationship given above, but also, to ensure stability, the members must be correctly positioned in the frame. In general, a frame built up as a series of triangles will be stable. For example, in Figure 2.1 frame (a) is stable and frame (b) is unstable (the right-hand panel is free to collapse, as indicated in Figure 2.1c).

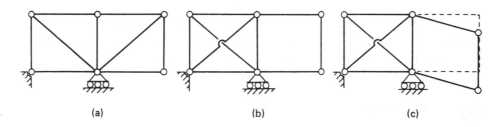

**Figure 2.1**

### (d) Method of Resolution at Joints

To determine the forces in all the members of a frame, it is, in general, best to use the method of *resolution at joints*. At any joint in a plane frame at which there are no more than *two* unknown forces, write down two equations of equilibrium obtained by resolving all forces at that joint in two mutually perpendicular directions. Solve these equations for the two unknown forces. Proceed systematically through the frame until all the forces are known. In the case of space frames, three equations of equilibrium may be written down; thus, joints with three unknown forces may be solved.

Usually forces will be resolved in the vertical and horizontal directions ($\Sigma V = 0$ and $\Sigma H = 0$). However, it may be convenient in some problems to resolve in two other mutually perpendicular directions. In many cases forces can be determined *by inspection* of the direction of members and forces acting at a particular joint.

External support reactions are usually determined before internal forces are calculated. However, depending on the geometry of the frame and the position of the supports, this may not always be necessary.

### (e) Method of Sections

To calculate the value of a force in only one or a few members of a frame, use the *method of sections*. Cut the frame, as indicated in Figure 2.2, by a section through the member under consideration and no more than two other members in which the member forces are unknown. Both parts of the structure can then be treated as structures in equilibrium and either part can then be solved by

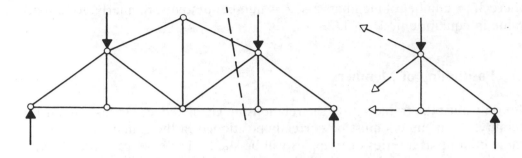

**Figure 2.2**  Frame cut for solution by method of sections

resolving forces or by taking moments of forces about a suitable point. The choice of whether to resolve or to take moments depends upon the geometry of the frame and will be illustrated in the examples in Section 2.4.

### (f)   Graphical Methods

Graphical methods of solution, although largely superseded by modern computer methods, may be useful to solve frames of complicated geometry. Commencing at a joint at which there are no more than two unknown forces, construct a force polygon for all the internal and external forces acting at that joint. The magnitude and the direction of the unknown forces can be scaled from the polygon. Proceed systematically through the frame, constructing a force polygon at each joint, to produce one composite force diagram for the whole structure.

## 2.3   Symbols, Units and Sign Conventions

$F$ = the internal force in a member (kN)
$V$ = the vertical component of reaction at a support (kN)
$H$ = horizontal component of reaction at a support (kN)
$R$ = the reaction at a support (kN)
$\theta$ = the acute angle between an inclined member and the horizontal

Tensile forces are assumed to be positive.
Clockwise moments are assumed to be positive.
Forces or components of forces acting vertically upward are assumed to be positive.
Forces or components of forces acting to the right are assumed to be positive.

In the various methods of analysis employed all unknown internal forces in the members of frames are assumed to be tensile. If the result of a calculation is a positive answer, then the initial assumption was correct, the member is in tension and it is a *tie*. If the answer is negative, the initial assumption was incorrect, the member must be in compression and it is a *strut*.

32

The assumed direction of action of unknown forces is indicated by open arrowheads on the diagrams of the frames.

The direction of action of known forces is indicated by solid arrowheads on the diagrams of the frames.

## 2.4 Worked Examples

### Example 2.1

The Warren truss shown in Figure 2.3 is composed of similar members all of which are 3 m long. Determine the forces in all the members due to a vertical load of 90 kN at G.

(Salford University)

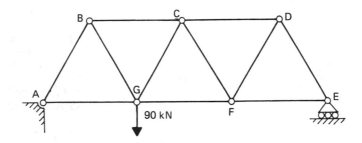

**Figure 2.3**

By inspection the frame is statically determinate; thus, the solution can proceed, using the conditions for statical equilibrium. Since it is necessary to determine the forces in all members, the method of resolution at joints will be used. Note that all members are the same length; thus, all internal angles are 60°.

It can be seen that there is no joint at which there are only two unknown forces; consequently, it is necessary to determine at least one reaction before commencing the resolution at joints.

*Solution 2.1*
(1)  To determine the support reactions
     Determine $V_E$ by taking moments about A:
     $(\Sigma M_A = 0)$

$$90 \times 3 - V_E \times 9 = 0$$

$$\therefore V_E = 30 \text{ kN}$$

Determine $V_A$ by taking moments about E:
$(\Sigma M_E = 0)$

$$V_A \times 9 - 90 \times 6 = 0$$

$$\therefore V_A = 60 \text{ kN}$$

Check $\Sigma V = 0$:
$$V_A + V_E = 30 + 60 = 90 \text{ kN} = \text{total load}$$

It is good practice and a check on arithmetic to calculate both reactions and to check that the total vertical reaction equals the total vertical load.

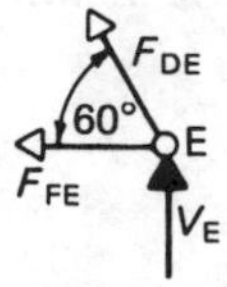

**Figure 2.4**

(2)  To determine member forces
    Resolving at E:
    $(\Sigma V = 0)$

$$F_{DE} \sin 60° + V_E = 0$$
$$F_{DE} \sin 60° + 30 = 0$$

$$\therefore \underline{F_{DE} = -34.64 \text{ kN}}$$

$(\Sigma H = 0)$

$$-F_{DE} \cos 60° - F_{FE} = 0$$
$$-(-34.64) \cos 60° - F_{FE} = 0$$

$$\therefore \underline{F_{FE} = +17.32 \text{ kN}}$$

Note that the above equations are set out assuming that all internal forces are tensile. Actual values are then substituted. A negative result implies that the force is compressive. Since the magnitude of $F_{DE}$ is now known, it is possible to resolve at joint D. Proceed systematically from joint to joint.

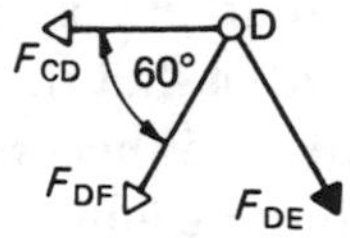

**Figure 2.5**

Resolving at D:
$(\Sigma V = 0)$

$$-F_{DF} \sin 60° - F_{DE} \sin 60° = 0$$
$$-F_{DF} \sin 60° - (-34.64) \sin 60° = 0$$

$$\therefore \underline{F_{DF} = +34.64 \text{ kN}}$$

$(\Sigma H = 0)$

$$-F_{CD} - F_{DF} \cos 60° + F_{DE} \cos 60° = 0$$
$$-F_{CD} - (+34.64) \cos 60° + (-34.64) \cos 60° = 0$$

$$\therefore \underline{F_{CD} = -34.64 \text{ kN}}$$

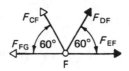

**Figure 2.6**

Resolving at F:
$(\Sigma V = 0)$

$$+F_{CF} \sin60° + F_{DF} \sin60° = 0$$
$$+F_{CF} \sin60° + (+34.64) \sin60° = 0$$

$$\therefore \underline{F_{CF} = -34.64 \text{ kN}}$$

$(\Sigma H = 0)$

$$-F_{FG} - F_{CF} \cos60° + F_{DF} \cos60° + F_{EF} = 0$$
$$-F_{FG} - (-34.64) \cos60° + (+34.64) \cos60° + (+17.32) = 0$$

$$\therefore \underline{F_{FG} = +51.96 \text{ kN}}$$

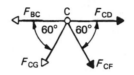

**Figure 2.7**

Resolving at C:
$(\Sigma V = 0)$

$$-F_{CG} \sin60° - F_{CF} \sin60° = 0$$
$$-F_{CG} \sin60° - (-34.64) \sin60° = 0$$

$$\therefore \underline{F_{CG} = +34.64 \text{ kN}}$$

$(\Sigma H = 0)$

$$-F_{BC} - F_{CG} \cos60° + F_{CF} \cos60° + F_{CD} = 0$$
$$-F_{BC} - (+34.64) \cos60° + (-34.64) \cos60° + (-34.64) = 0$$

$$\therefore \underline{F_{BC} = -69.28 \text{ kN}}$$

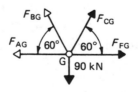

**Figure 2.8**

Resolving at G:
$(\Sigma V = 0)$

$$+F_{BG} \sin60° + F_{CG} \sin60° - 90 = 0$$
$$+F_{BG} \sin60° + (+34.64) \sin60° - 90 = 0$$

$$\therefore \underline{F_{BG} = +69.28 \text{ kN}}$$

35

$(\Sigma H = 0)$

$$-F_{AG} - F_{BG}\cos60° + F_{CG}\cos60° + F_{FG} = 0$$
$$-F_{AG} - (+69.28)\cos60° + (+34.64)\cos60° + (+51.96) = 0$$

$$\therefore \underline{F_{AG} = +34.64 \text{ kN}}$$

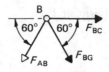

**Figure 2.9**

Resolving at B:
$(\Sigma V = 0)$

$$-F_{AB}\sin60° - F_{BG}\sin60° = 0$$
$$-F_{AB}\sin60° - (+69.28)\sin60° = 0$$

$$\therefore \underline{F_{AB} = -69.28 \text{ kN}}$$

**Figure 2.10**

Check by resolving at A:

$$\Sigma V = +F_{AB}\sin60° + V_A = (-69.28)\sin60° + 60$$
$$= 0$$

$$\therefore \text{ correct}$$

$$\Sigma H = +F_{AB}\cos60° + F_{AG} = (-69.28)\cos60° + (+34.68)$$
$$= 0$$

$$\therefore \text{ correct}$$

**Resolving at A provides a useful arithmetical check.**

*Summary*

| Member | Force | Member | Force | Member | Force |
|--------|-------|--------|-------|--------|-------|
| AB | −69.28 | AG | +34.64 | BC | −69.28 |
| BG | +69.28 | CD | −34.64 | CF | −34.64 |
| CG | +34.64 | DE | −34.64 | DF | +34.64 |
| EF | +17.32 | FG | +51.96 | | |

It is useful to summarise the results, either in a table as shown above or on a diagram of the frame (Figure 2.11). Such a diagram simplifies future reference to, and checking of, the solution. It also makes it easier for the examiner to check your solution.

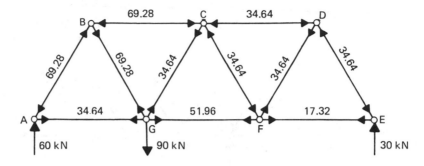

**Figure 2.11**   Answers to Example 2.1 shown on a diagram of the frame

A diagram such as that above should be developed progressively as the analysis proceeds. The magnitude and true direction of each force should be plotted immediately they have been determined. Ensure that you do this in all subsequent problems.

## Example 2.2

The roof truss shown in Figure 2.12 has all of its members aligned at 0°, 30° or 60° to the horizontal. It is acted on by the wind forces shown, which are perpendicular to the rafters. Determine all the member forces caused by this loading. Use an initial inspection to identify any obviously unloaded members. Summarise the member forces on a sketch of the truss.

(Sheffield University)

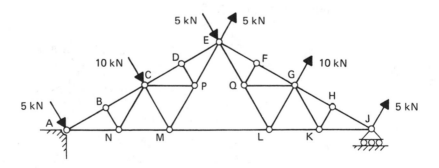

**Figure 2.12**

(Note that all wind forces are at 90° to the rafters AE and EJ and are at 60° to the horizontal.)

It is always advisable to study a frame to see whether the force in any member can be determined by inspection. In this question you are specifically asked to do so. Your compliance with this should be made clear in the solution.

### *Solution 2.2*

By inspection at joints B, D, F and H the forces in members BN, DP, FQ and HK equal zero. Figure 2.13 shows these members marked accordingly.

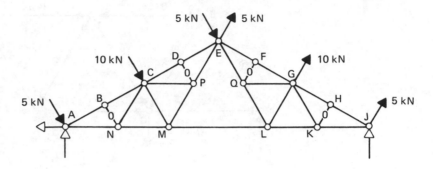

**Figure 2.13**

**If you have difficulty in seeing that the force is zero in these members, try resolving in a direction at right angles to the rafters AE and EJ.**

The reactions at the supports have been shown by open arrows in Figure 2.13, since their magnitude is not known. The vertical components of reaction at both A and J have been assumed to be acting upwards (positive), whereas the horizontal component of reaction at A has been shown as acting to the left (negative). This direction for $H_A$ is determined by inspection, since all the loads have horizontal components acting to the right.

By inspection at joints N, P, Q and K (**in Figure 2.13**) the forces in members NC, PC, QG and KG equal zero. **These are shown in Figure 2.14. If you are not sure about this, then resolve at joint N in a direction perpendicular to AJ and at P in a direction perpendicular to EM. Similarly at joints K and Q.**

By inspection at joints C and G (**in Figure 2.14**) the forces in members CM and GL are equal to 10 kN compressive and 10 kN tensile, respectively. **These are shown in Figure 2.15.**

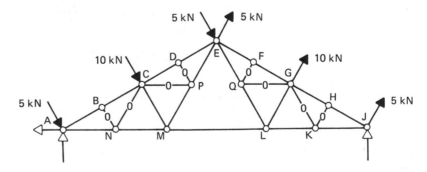

**Figure 2.14**

By inspection at joint M and subsequently at P (**in Figure 2.15**) the forces in members MP and PE are equal to 10 kN tensile. Similarly, by inspection at joints L and Q the forces in members LQ and QE are equal to 10 kN compressive. **These are shown in Figure 2.16.**

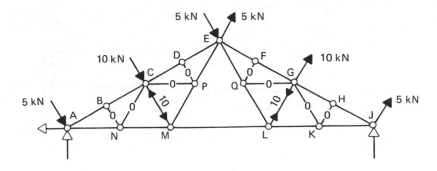

**Figure 2.15**

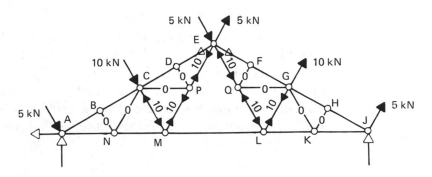

**Figure 2.16**

By resolving at E:

($\Sigma V = 0$) **Note that in drafting the following equations all member forces are initially assumed to be tensile. Any known values are then substituted in the second line of the calculation.**

$$-F_{DE} \sin30° - F_{EP} \sin60° - F_{EQ} \sin60° - F_{EF} \sin30° + 5 \sin60° - 5 \sin60° = 0$$
$$-F_{DE} \sin30° - (+10.00) \sin60° - (-10.00) \sin60° - F_{EF} \sin30° = 0$$

$$\therefore -0.5F_{DE} - 0.5F_{EF} = 0$$

$$\therefore \underline{F_{DE} = -F_{EF}}$$

($\Sigma H = 0$)
$$-F_{DE} \cos30° - F_{EP} \cos60° + F_{EQ} \cos60° + F_{EF} \cos30° + 5 \cos60°$$
$$+ 5 \cos60° = 0$$
$$-F_{DE} \cos30° - (+10.00) \cos60° + (-10.00) \cos60° + F_{EF} \cos30°$$
$$+ 10 \cos60° = 0$$
$$-0.866F_{DE} + 0.866F_{EF} = 5.00$$
$$-0.866(-F_{EF}) + 0.866F_{EF} = 5.00$$
$$1.732F_{EF} = 5.00$$

$$\therefore \underline{F_{EF} = +2.89 \text{ kN}}$$

and

$$\underline{F_{DE} = -2.89 \text{ kN}}$$

**These forces are shown on Figure 2.17.**

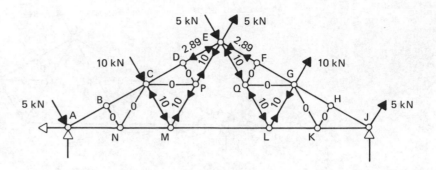

**Figure 2.17**

By inspection at joints D, C and B in sequence, the forces in members CD, BC and AB are equal to 2.89 kN compressive. Similarly, by inspection at F, G and H the forces in members FG, GH and HJ are equal to 2.89 kN tensile. **These forces are shown on Figure 2.18. Note that $V_J$ is now shown acting downwards. This direction of action is determined by inspection at joint J (i.e. by considering the vertical components of all forces acting at that joint).**

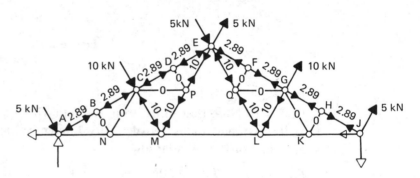

**Figure 2.18**

Resolving horizontally at joint J:
($\Sigma H = 0$)

$$-F_{HJ} \cos 30° - F_{JK} + 5 \cos 60° = 0$$
$$-(+2.89) \cos 30° - F_{JK} + 5 \cos 60° = 0$$

$$\therefore \underline{F_{JK} = 0.0}$$

By inspection at joint K:

$$\underline{F_{KL} = 0.0}$$

**These results are shown on Figure 2.19.**

Resolving horizontally at joint L:
($\Sigma H = 0$)

$$+F_{KL} + F_{GL} \cos 60° - F_{LQ} \cos 60° - F_{LM} = 0$$
$$+(0.0) + (+10.0) \cos 60° - (-10.0) \cos 60° - F_{LM} = 0$$

$$\therefore \underline{F_{LM} = +10.00 \text{ kN}}$$

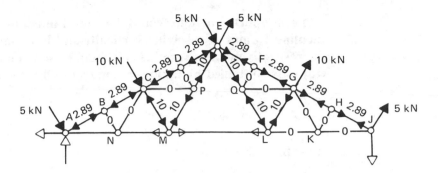

**Figure 2.19**

Resolving horizontally at joint M:
($\Sigma H = 0$)

$$+F_{LM} + F_{MP}\cos 60° - F_{CM}\cos 60° - F_{MN} = 0$$
$$+(+10.0) + (+10.0)\cos 60° - (-10.0)\cos 60° - F_{MN} = 0$$

$$\therefore F_{MN} = +20.00 \text{ kN}$$

By inspection at joint N (**resolving horizontally**):

$$F_{AN} = +20.00 \text{ kN}$$

**These forces are shown in Figure 2.20, which shows the forces in all members of the frame.**

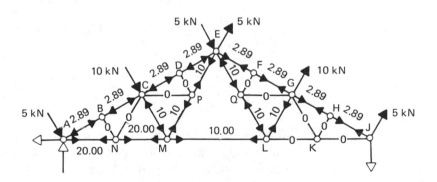

**Figure 2.20**

In the solution just completed, eight diagrams have been used. This has been done for the benefit of the reader and to make it easier to follow the steps in the analysis. In practice, it is anticipated that the student would construct one figure only (2.20), inserting the values of the forces as he/she proceeds through the analysis.

Note that although the frame in this example looks complex and might at first sight suggest a time-consuming solution, only five calculations for member forces are required, the majority of the analysis being by inspection. This emphasises the importance of a careful appraisal of all problems before selecting which to attempt in an examination.

The support reactions could be determined by resolving vertically at J to calculate $V_J$ and by resolving vertically and horizontally at A to calculate $V_A$ and $H_A$, respectively. An overall check could then be made by comparing these values with those obtained by taking moments of all external forces about one support.

## Example 2.3

Explain why the pin-jointed frame shown in Figure 2.21 is statically determinate. Determine the forces in the members of this frame when the loads shown are applied. Use any non-graphical method for your analysis.

(Manchester University)

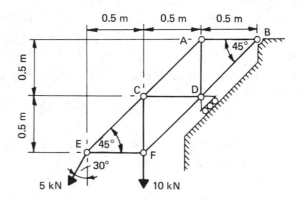

**Figure 2.21**

## *Solution 2.3*

(1) To answer the first part of the question

Support D, being a roller, provides one reaction, $R_D$, which will act at right angles to the direction of movement of the rollers. Support B, being a pin, provides a reaction whose direction is unknown but can be calculated if the value of two components, $H_B$ (horizontally) and $V_B$ (vertically), can be determined. Thus, there are three support reactions whose directions are known but whose magnitudes are not known. These three unknowns may be determined by use of the three unique equations for statical equilibrium ($\Sigma M = 0$, $\Sigma V = 0$, $\Sigma H = 0$). The frame is thus *externally statically determinate*.

For internal statical determinacy $M = 2J - 3$. In this example $M = 9$ and $J = 6$.

$$\therefore 2J - 3 = (2 \times 6) - 3 = 9 \quad (\text{which} = M)$$

Thus, the frame has sufficient (but no more than sufficient) members, and since inspection shows that the members are correctly positioned, then the frame is also *internally statically determinate*.

42

(2) To determine member forces

**Since the values of the forces in *all* members are required, the method of resolution at joints will be used. Note that in this example it is possible to start resolving at joint E without first calculating the value of the reactions.**

Resolving at E:

$(\Sigma V = 0)$

$$+F_{CE} \sin45° - 5 \sin60° = 0$$

$$\therefore \underline{F_{CE} = + 6.12 \text{ kN}}$$

$(\Sigma H = 0)$

$$F_{CE} \cos45° + F_{EF} - 5 \cos60° = 0$$
$$(+6.12) \cos45° + F_{EF} - 5 \cos60° = 0$$

$$\therefore \underline{F_{EF} = -1.83 \text{ kN}}$$

Resolving at F:

$(\Sigma H = 0)$

$$F_{DF} \cos45° - F_{EF} = 0$$
$$F_{DF} \cos45° - (-1.83) = 0$$

$$\therefore \underline{F_{DF} = -2.59 \text{ kN}}$$

$(\Sigma V = 0)$

$$F_{CF} + F_{DF} \sin45° - 10 = 0$$
$$F_{CF} + (-2.59) \sin45° - 10 = 0$$

$$\therefore \underline{F_{CF} = +11.83 \text{ kN}}$$

Resolving at C:

$(\Sigma V = 0)$

$$F_{AC} \sin45° - F_{CE} \sin45° - F_{CF} = 0$$
$$F_{AC} \sin45° - (+6.12) \sin45° - (+11.83) = 0$$

$$\therefore \underline{F_{AC} = +22.85 \text{ kN}}$$

$(\Sigma H = 0)$

$$F_{AC} \cos45° + F_{CD} - F_{CE} \cos45° = 0$$
$$(+22.85) \cos45° + F_{CD} - (+6.12) \cos45° = 0$$

$$\therefore \underline{F_{CD} = -11.83 \text{ kN}}$$

Resolving at A:

$(\Sigma V = 0)$

$$-F_{AD} - F_{AC} \sin45° = 0$$
$$-F_{AD} - (+22.85) \sin45° = 0$$

$$\therefore \underline{F_{AD} = -16.16 \text{ kN}}$$

$(\Sigma H = 0)$

$$F_{AB} - F_{AC} \cos45° = 0$$
$$F_{AB} - (+22.85) \cos45° = 0$$

$$\therefore \underline{F_{AB} = +16.16 \text{ kN}}$$

Resolving at D (parallel to BDF):

At this joint, there are now two forces ($F_{BD}$ and $R_D$) of unknown magnitude acting parallel to and at right angles to BDF, respectively. It is more convenient to resolve in these two directions, since resolving in the vertical and horizontal directions would result in the necessity to solve simultaneous equations.

Resolving parallel to BDF:

$$F_{BD} + F_{AD} \cos45° - F_{CD} \cos45° - F_{DF} = 0$$
$$F_{BD} + (-16.16) \cos45° - (-11.83) \cos45° - (-2.59) = 0$$

$$\therefore \underline{F_{BD} = +0.47 \text{ kN}}$$

Resolving perpendicular to BDF:

$$+F_{AD} \sin45° + F_{CD} \sin45° + R_D = 0$$
$$(-16.16) \sin45° + (-11.83) \sin45° + R_D = 0$$

$$\therefore \underline{R_D = +19.79 \text{ kN}}$$

The question did not ask for the value of the reactions, but it is useful to determine them, since this will enable an arithmetical check to be made.

Resolving at B:
($\Sigma H = 0$)

$$H_B - F_{AB} - F_{BD} \cos45° = 0$$
$$H_B - (+16.16) - (+0.47) \cos45° = 0$$

$$\therefore \underline{H_B = +16.49 \text{ kN}}$$

($\Sigma V = 0$)

$$V_B - F_{BD} \sin45° = 0$$
$$V_B - (+0.47) \sin45° = 0$$

$$\therefore \underline{V_B = +0.33 \text{ kN}}$$

Check by considering the whole frame and resolving all external forces:

$$\Sigma H = H_B - R_D \cos45° - 5 \cos60°$$
$$= (+16.49) - (+19.79) \cos45° - 5 \cos60°$$
$$= 0 \quad \text{(correct)}$$
$$\Sigma V = +R_D \sin45° + V_B - 10 - 5 \sin60°$$
$$= +(+19.79) \sin45° + (+0.33) - 10 - 5 \sin60°$$
$$= 0 \quad \text{(correct)}$$

*Summary*
**The magnitude and direction of all internal forces and of the reactions are shown on the diagram of the complete frame (Figure 2.22).**

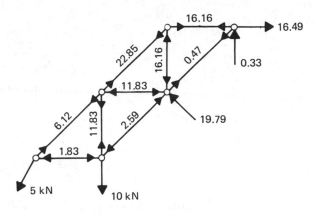

Figure 2.22

## Example 2.4

The pin-jointed frame shown in Figure 2.23 is pinned to foundations at A and E. Calculate the support reactions and the forces in members BD and DF.

(Birmingham University)

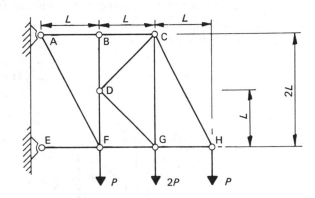

Figure 2.23

**Support A is a pin providing a reaction whose direction of action is initially unknown. It will, therefore, be necessary to determine values of two components ($H_A$ and $V_A$). E is also a pinned joint, but since there is only one member connected to that joint, then the direction of the reaction is known (it must be collinear with EF, which is horizontal). Thus, it is necessary to determine the value of a total of three external forces ($H_A$, $V_A$ and $H_E$).**

*Solution 2.4*

(1)  To determine the support reactions
     Taking moments about A:
     ($\Sigma M_A = 0$)
     $$+(P \times L) + (2P \times 2L) + (P \times 3L) - (H_E \times 2L) = 0$$
     $$\therefore \underline{H_E = +4P}$$

45

$(\Sigma H = 0)$

$$H_A + H_E = 0$$
$$H_A + (+4P) = 0$$

$$\therefore \underline{H_A = -4P} \text{ (4P acting to the left)}$$

$(\Sigma V = 0)$

$$+V_A - P - 2P - P = 0$$

$$\therefore \underline{V_A = +4P} \text{ (4P acting upwards)}$$

(2) To determine forces in BD and DF
By inspection:

$$\underline{F_{BD} = 0}$$

**If in doubt, try resolving vertically at joint B.**

To determine $F_{DF}$ use the method of sections, cutting the frame by a section $X$–$X$ through AB, DF and FG as shown in Figure 2.24.
Consider the right-hand part of the frame, as in Figure 2.24.
Resolving vertically:
$(\Sigma V = 0)$

$$-F_{DF} - 2P - P = 0$$
$$\therefore \underline{F_{DF} = -3P} \text{ (i.e. 3P compression)}$$

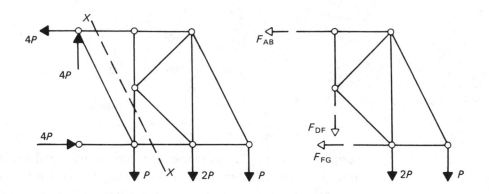

**Figure 2.24**

**Note that although the force in member DF could have been obtained by resolving at joints, first at joint A and then at joint F, the method of sections provides a quicker solution.**

*Summary*

$$H_E = 4P \text{ to the right}$$
$$H_A = 4P \text{ to the left; } V_A = 4P \text{ upwards}$$
$$F_{BD} = 0$$
$$F_{DF} = 3P \text{ compression}$$

**Example 2.5**

Figure 2.25 shows the details of a loaded pin-jointed plane frame. Determine the nature and magnitude of the forces in members BC, CH, GH, DE and EF.

(University of Hertfordshire)

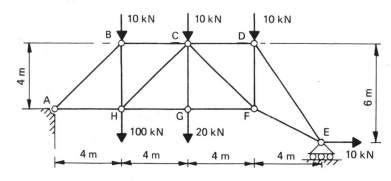

**Figure 2.25**

This question suggests the use of the method of sections, and inspection shows that a single section through members BC, CH and GH will enable three of the unknowns to be determined. It will first be necessary to determine the values of the support reactions.

*Solution 2.5*

(1)  To determine the support reactions
Taking moments about A:
$(\Sigma M_A = 0)$

$$+(10 \times 4) + (10 \times 8) + (10 \times 12) + (100 \times 4) + (20 \times 8)$$
$$- (10 \times (6 - 4)) - (V_E \times 16) = 0$$

$$\therefore \underline{V_E = +48.75 \text{ kN}}$$

Resolving horizontally:
$(\Sigma H = 0)$

$$H_A + 10 = 0$$

$$\therefore \underline{H_A = -10 \text{ kN}}$$

Resolving vertically:
$(\Sigma V = 0)$

$$V_A + V_E - 10 - 10 - 10 - 100 - 20 = 0$$
$$V_A + (+48.75) - 10 - 10 - 10 - 100 - 20 = 0$$

$$\therefore \underline{V_A = +101.25 \text{ kN}}$$

Check by taking moments about E:
$(\Sigma M_E = 0)$

$$+(H_A \times (6 - 4)) + (V_A \times 16) - (10 \times 12) - (10 \times 8) - (10 \times 4)$$
$$- (100 \times 12) - (20 \times 8) = (-10 \times 2) + (+101.25 \times 16) - 1600$$
$$= 0 \quad \text{(correct)}$$

(2) To determine forces in BC, CH and GH

Cut the frame by a section $X$–$X$ as in Figure 2.26.

Considering the left-hand part:

Resolving vertically,

($\Sigma V = 0$)

$$+F_{CH} \sin 45° + V_A - 10 - 100 = 0$$
$$+F_{CH} \sin 45° + (+101.25) - 10 - 100 = 0$$

$$\therefore \underline{F_{CH} = +12.37 \text{ kN}}$$

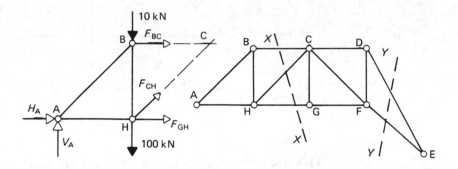

**Figure 2.26**

Taking moments about joint H,

($\Sigma M_H = 0$)

$$+(F_{BC} \times 4) + (V_A \times 4) = 0$$
$$+(F_{BC} \times 4) + (+101.25 \times 4) = 0$$

$$\therefore \underline{F_{BC} = -101.25 \text{ kN}}$$

Taking moments about joint C,

($\Sigma M_C = 0$)

$$+(V_A \times 8) - (H_A \times 4) - (10 \times 4) - (100 \times 4) - (F_{GH} \times 4) = 0$$
$$+(+101.25 \times 8) - (-10 \times 4) - (10 \times 4) - (100 \times 4) - (F_{GH} \times 4) = 0$$

$$\therefore \underline{F_{GH} = +102.5 \text{ kN}}$$

(3) To determine forces in members DE and EF

Resolving all forces at joint E (**see Figure 2.27**):

($\Sigma V = 0$)

$$+F_{DE} \sin\theta_1 + F_{EF} \sin\theta_2 + V_E = 0$$

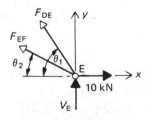

**Figure 2.27**

where, from the geometry of the frame,

$$\sin\theta_1 = 6/(6^2 + 4^2)^{1/2} = 0.8320$$
$$\sin\theta_2 = 2/(2^2 + 4^2)^{1/2} = 0.4472$$

$$\therefore +F_{DE} \times 0.8320 + F_{EF} \times 0.4472 + (+48.75) = 0$$

Rearranging,

$$0.8320F_{DE} + 0.4472F_{EF} = -48.75 \qquad (1)$$

$(\Sigma H = 0)$

$$-F_{DE}\cos\theta_1 - F_{EF}\cos\theta_2 + 10 = 0$$

where

$$\cos\theta_1 = 4/(6^2 + 4^2)^{1/2} = 0.5547$$
$$\cos\theta_2 = 4/(2^2 + 4^2)^{1/2} = 0.8944$$

$$\therefore -F_{DE} \times 0.5547 - F_{EF} \times 0.8944 + 10 = 0$$

Rearranging,

$$-0.5547F_{DE} - 0.8944F_{EF} = -10 \qquad (2)$$

Equation (1) multiplied by $0.5547/0.8320$ becomes

$$0.5547F_{DE} + 0.2981F_{EF} = -32.5019$$

Equation (2)    $\underline{-\,0.5547F_{DE} - 0.8944F_{EF} = -10.0000}$

Adding,    $-0.5963F_{EF} = -42.5019$

$$\therefore \underline{F_{EF} = +71.28 \text{ kN}}$$

Substituting in Equation (1),

$$0.8320F_{DE} + 0.4472(+71.28) = -48.75$$

$$\therefore \underline{F_{DE} = -96.90 \text{ kN}}$$

*Summary*

| Member | Force | Type |
|--------|-------|------|
| BC | 101.25 kN | compressive |
| CH | 12.37 kN | tensile |
| GH | 102.50 kN | tensile |
| DE | 96.90 kN | compressive |
| EF | 71.28 kN | tensile |

**Note that if the frame had been cut by a section *Y–Y* (see Figure 2.26), the right-hand part of the frame would consist of only one joint and two cut members. Thus, the solution is a simple resolution of forces at joint E.**

**The geometry of this frame is such that, when resolving at joint E, it is necessary to work to four decimal places in order to obtain reasonably accurate**

values for the forces $F_{DE}$ and $F_{EF}$. An efficient, and quicker, method of setting down the simultaneous equations to give accurate values for the forces is by the use of *tension coefficients*. The tension coefficient $t_{DE}$ for member DE is the force in that member divided by the length of the member. That is,

$$t_{DE} = F_{DE}/(6^2 + 4^2)^{1/2}$$

The vertical component of $F_{DE} = t_{DE} \times$ the projection on to the vertical axis of the length of the member DE. Thus,

$$\text{vertical component of } F_{DE} = t_{DE} \times 6$$

and, similarly,

$$\text{horizontal component of } F_{DE} = t_{DE} \times 4$$

The components of force in member EF can be expressed in a similar manner. Hence, the resolving of forces at joint E could be set out as follows:

Resolving at E:
$(\Sigma V = 0)$

$$+6t_{DE} + 2t_{EF} + 48.75 = 0$$

$(\Sigma H = 0)$

$$-4t_{DE} - 4t_{EF} + 10 = 0$$

Solving these equations gives

$$t_{DE} = -13.437$$

Thus,

$$F_{DE} = -13.437 \times (6^2 + 4^2)^{1/2}$$

$$\therefore \underline{F_{DE} = -96.90 \text{ kN}}$$

$$t_{EF} = +15.938$$

Thus,

$$F_{EF} = +15.938 \times (2^2 + 4^2)^{1/2}$$

$$\therefore \underline{F_{EF} = +71.28 \text{ kN}}$$

## Example 2.6

Figure 2.28 shows a pin-jointed framework supporting a vertical and a horizontal load. Determine the three reactive components needed to maintain overall equilibrium and the magnitude of the forces in members FD, FE, FG, and CE, stating whether the forces are tensile or compressive.

(Nottingham University)

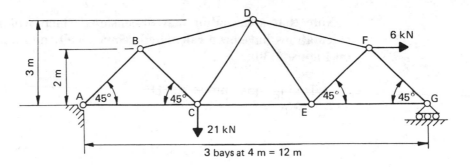

**Figure 2.28**

## Solution 2.6

(1) To determine support reactions
Taking moments about A:
$(\Sigma M_A = 0)$

$$+(21 \times 4) + (6 \cdot \times 2) - (V_G \times 12) = 0$$

$$\therefore \underline{V_G = +8.00 \text{ kN}}$$

$(\Sigma V = 0)$

$$+V_A + V_G - 21 = 0$$
$$V_A + (+8.00) - 21 = 0$$

$$\therefore \underline{V_A = +13.00 \text{ kN}}$$

$(\Sigma H = 0)$

$$+H_A + 6 = 0$$

$$\therefore \underline{H_A = -6.00 \text{ kN}}$$

Check by taking moments about G:
$(\Sigma M_G = 0)$
$$+(V_A \times 12) + (6 \times 2) - (21 \times 8) = +(+13.00 \times 12) + 12 - 168$$
$$= 0 \quad \text{(correct)}$$

(2) To determine member forces
Cut the frame at section $X$–$X$ as in Figure 2.29 and consider the equilibrium of the right-hand part.

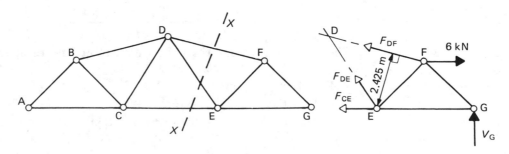

**Figure 2.29**

51

Note that some additional dimensions which will be needed in subsequent calculations have been calculated geometrically and marked on this diagram and on Figure 2.30.

Taking moments about D:
($\Sigma M_D = 0$)

$$+(F_{CE} \times 3) - (6 \times 1) - (V_G \times 6) = 0$$
$$+(F_{CE} \times 3) - (6 \times 1) - (+8 \times 6) = 0$$

$$\therefore \underline{F_{CE} = +18.00 \text{ kN}} \quad \text{(tensile)}$$

Taking moments about E:
($\Sigma M_E = 0$)

$$-(F_{DF} \times 2.425) - (V_G \times 4) + (6 \times 2) = 0$$
$$-(F_{DF} \times 2.425) - (+8 \times 4) + (6 \times 2) = 0$$

$$\therefore \underline{F_{DF} = -8.25 \text{ kN}} \quad \text{(compressive)}$$

Cut the frame at section $Y$–$Y$ and consider the right-hand part as in Figure 2.30.

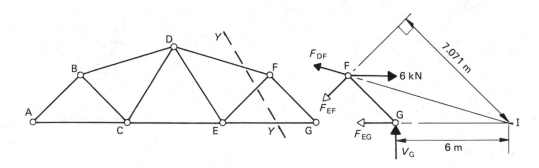

**Figure 2.30**

Taking moments about I:

**I is the intersection point of members DF and EG extended as shown.**

($\Sigma M_I = 0$)

$$-(F_{EF} \times 7.071) + (6 \times 2) + (V_G \times 6) = 0$$
$$-(F_{EF} \times 7.071) + (6 \times 2) + (+8 \times 6) = 0$$

$$\therefore \underline{F_{EF} = +8.48 \text{ kN}} \quad \text{(tensile)}$$

**The force in member FG is best determined by resolving at joint G.**

Resolving at joint G:
($\Sigma V = 0$)

$$+F_{FG} \sin 45° + V_G = 0$$
$$+F_{FG} \sin 45° + (+8) = 0$$

$$\therefore \underline{F_{FG} = -11.31 \text{ kN}} \quad \text{(compressive)}$$

$$V_A = 13.00 \text{ kN upwards}$$
$$H_A = 6.00 \text{ kN to the left}$$
$$V_G = 8.00 \text{ kN upwards}$$

| Member | Force | Type |
|--------|-------|------|
| FD | 8.25 kN | compressive |
| FE | 8.48 kN | tensile |
| FG | 11.31 kN | compressive |
| CE | 18.00 kN | tensile |

## Example 2.7

Using the method of sections, or otherwise, determine the forces in the members GF and HF for the truss shown in Figure 2.31. Hence determine the magnitude of the forces in the members meeting at joint G.

(Liverpool University)

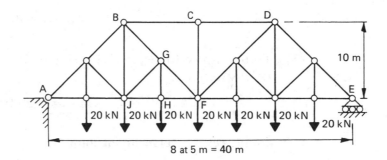

**Figure 2.31**

## *Solution 2.7*

(1)  To determine the support reactions

**The loading is symmetrical; thus, the two vertical reactions will be equal and of value one-half of the total load. There are no horizontal loads; consequently, there is no horizontal reaction at the pinned joint A.**

$$\text{Total load} = 7 \times 20 = 140 \text{ kN}$$

$$\therefore V_A = V_E = 70 \text{ kN}$$

(2)  To determine forces in GF and HF
Cut the frame by a section *X–X* and consider the left-hand part as in Figure 2.32.

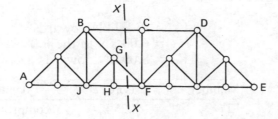
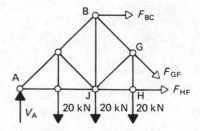

**Figure 2.32**

Resolving vertically:
$(\Sigma V = 0)$

$$+V_A - 20 - 20 - 20 - F_{GF}\sin45° = 0$$
$$+70 - 20 - 20 - 20 - F_{GF}\sin45° = 0$$

$$\therefore F_{GF} = +14.14 \text{ kN}$$

Taking moments about B:
$(\Sigma M_B = 0)$

$$+(V_A \times 10) + (20 \times 5) - (20 \times 5) - (F_{HF} \times 10) = 0$$
$$+(+70 \times 10) + (20 \times 5) - (20 \times 5) - (F_{HF} \times 10) = 0$$

$$\therefore \underline{F_{HF} = +70.00 \text{ kN}}$$

(3) To determine forces in GH, GJ and BG
By inspection,

$$\underline{F_{GH} = +20.00 \text{ kN}}$$

**If in doubt, resolve vertically at joint H.**

Resolving perpendicular to BGF at joint G:

$$+F_{GJ} + F_{GH}\cos45° = 0$$
$$+F_{GJ} + (+20.00)\cos45° = 0$$

$$\therefore \underline{F_{GJ} = -14.14 \text{ kN}}$$

**Note that member GJ is at right angles to BGF.**

Resolving parallel to BGF at joint G:

$$+F_{BG} - F_{GF} - F_{GH}\sin45° = 0$$
$$+F_{BG} - (+14.14) - (+20.00)\sin45° = 0$$

$$\therefore \underline{F_{BG} = +28.28 \text{ kN}}$$

*Summary*

| Member | Force | Type |
|--------|-------|------|
| GF | 14.14 kN | tensile |
| HF | 70.00 kN | tensile |
| GH | 20.00 kN | tensile |
| GJ | 14.14 kN | compressive |
| BG | 28.28 kN | tensile |

## Example 2.8

The pin-jointed frame shown in Figure 2.33 is pinned to foundations at D and G. There are two independent pins at E and F which are not connected, but there is only one pin at B. Determine the support reactions and the force in each member of the frame for the loading shown in the figure. Indicate clearly which bar forces are tensile and which are compressive.

(Birmingham University)

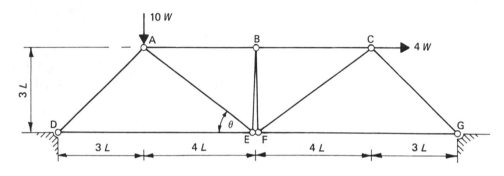

**Figure 2.33**

**This frame is essentially a three-pinned structure, as dealt with in Chapter 1. An additional equilibrium equation, based on the fact that the moment at the central pin at joint B is zero, will be needed to determine the reactions.**

*Solution 2.8*

(1)  To determine the support reactions
     Considering the whole frame:
     Taking moments about G,
     $(\Sigma M_G = 0)$

$$+(V_D \times 14L) + (4W \times 3L) - (10W \times 11L) = 0$$

$$\therefore \underline{V_D = +7.00W}$$

$(\Sigma V = 0)$

$$+V_D + V_G - 10W = 0$$

$$\therefore \underline{V_G = +3.00W}$$

55

Considering the right-hand part only:
Taking moments about B,
$(\Sigma M_B = 0)$

$$-(H_G \times 3L) - (V_G \times 7L) = 0$$
$$-(H_G \times 3L) - (+3W \times 7L) = 0$$

$$\therefore \underline{H_G = -7.00W} \quad \text{(i.e. to the left)}$$

Considering the whole frame:
$(\Sigma H = 0)$

$$+H_D + H_G + 4W = 0$$
$$+H_D + (-7.00W) + 4W = 0$$

$$\therefore \underline{H_D = +3.00W} \quad \text{(to the right)}$$

(2)  To determine the member forces
Resolving at D:
$(\Sigma V = 0)$

$$+F_{AD} \sin 45° + V_D = 0$$
$$+F_{AD} \sin 45° + (+7.00W) = 0$$

$$\therefore \underline{F_{AD} = -9.90W}$$

$(\Sigma H = 0)$

$$+F_{DE} + F_{AD} \cos 45° + H_D = 0$$
$$+F_{DE} + (-9.90W) \cos 45° + (+3.00W) = 0$$

$$\therefore \underline{F_{DE} = +4.00W}$$

Resolving at E:
$(\Sigma H = 0)$

$$-F_{DE} - F_{AE} \cos\theta = 0$$

**where $\theta = \tan^{-1} \frac{3}{4}$ , as in Figure 2.33. $\sin\theta = \frac{3}{5}$ and $\cos\theta = \frac{4}{5}$.**

$$\therefore -(+4.00W) - F_{AE} \times \tfrac{4}{5} = 0$$

$$\therefore \underline{F_{AE} = -5.00W}$$

$(\Sigma V = 0)$

$$+F_{BE} + F_{AE} \sin\theta = 0$$
$$+F_{BE} + (-5.00W)\tfrac{3}{5} = 0$$

$$\therefore \underline{F_{BE} = +3.00W}$$

Resolving at A:
$(\Sigma H = 0)$

$$+F_{AB} + F_{AE} \cos\theta - F_{AD} \cos 45° = 0$$
$$F_{AB} + (-5.00W)\tfrac{4}{5} - (-9.90W) \cos 45° = 0$$

$$\therefore \underline{F_{AB} = -3.00W}$$

Resolving at G:
($\Sigma V = 0$)

$$+F_{CG} \sin 45° + V_G = 0$$
$$+F_{CG} \sin 45° + (+3.00W) = 0$$
$$\therefore \underline{F_{CG} = -4.24W}$$

($\Sigma H = 0$)

$$-F_{FG} - F_{GC} \cos 45° + H_G = 0$$
$$-F_{FG} - (-4.24W) \cos 45° + (-7.00W) = 0$$
$$\therefore \underline{F_{FG} = -4.00W}$$

Resolving at F:
($\Sigma H = 0$)

$$+F_{FG} + F_{CF} \cos\theta = 0$$
$$+(-4.00W) + F_{CF} \times \tfrac{4}{5} = 0$$
$$\therefore \underline{F_{CF} = +5.00W}$$

($\Sigma V = 0$)

$$+F_{BF} + F_{CF} \sin\theta = 0$$
$$+F_{BF} + (+5.00W)\tfrac{3}{5} = 0$$
$$\therefore \underline{F_{BF} = -3.00W}$$

Resolving at C:
($\Sigma H = 0$)

$$+4.00W + F_{CG} \cos 45° - F_{CF} \cos\theta - F_{BC} = 0$$
$$+4.00W + (-4.24W) \cos 45° - (+5.00W)\tfrac{4}{5} - F_{BC} = 0$$
$$\therefore \underline{F_{BC} = -3.00W}$$

Check at B:

$$\Sigma H = 0 \quad \text{and} \quad \Sigma V = 0 \quad \text{(correct)}$$

*Summary*

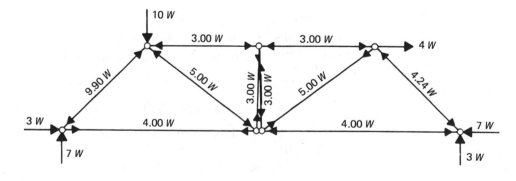

**Figure 2.34**

57

**Example 2.9**

Determine the force in each leg of the pin-jointed tripod, shown in Figure 2.35, using the coordinate system shown. A, B and C are pinned to a horizontal base and D is 4 m above the base.

(University of Westminster)

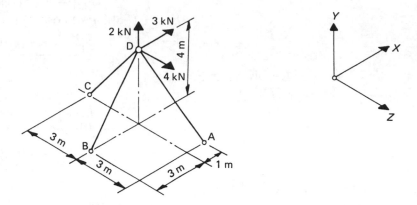

**Figure 2.35**

**This example is included to show how the use of *tension coefficients* (as defined in Example 2.5) may be used to set up the sets of simultaneous equations required for the solution of a space frame. It is useful, as a first step to draw an orthographic plan and an elevation (see Figure 2.36), so that dimensions projected onto the X, Y and Z axes are easily determined. In this example it is only necessary to resolve at one joint. However, the method may be applied to more complicated frames by resolving at a greater number of joints.**

*Solution 2.9*

(1)  Resolving at joint D
$(\Sigma X = 0)$

$$+3.00 + (t_{AD} \times x_{AD}) + (t_{BD} \times x_{BD}) + (t_{CD} \times x_{CD}) = 0$$

where $t_{AD}$ is the *tension coefficient* for member AD (i.e. the force in member AD divided by the length of member AD) and $x_{AD}$ is the projection on the X axis of the length AD. Similarly for $t_{BD}$ and $t_{CD}$.

By inspection (from Figure 2.36),

$$x_{AD} = +1.0; \; x_{BD} = -3.0; \; x_{CD} = 0.0$$

Therefore, the above equation becomes
$(\Sigma X = 0)$

$$+3.00 + (t_{AD} \times + 1.0) + (t_{BD} \times -3.0) + (t_{CD} \times 0) = 0$$
$$+1.0t_{AD} - 3.0t_{BD} + 3.0 = 0 \qquad (1)$$

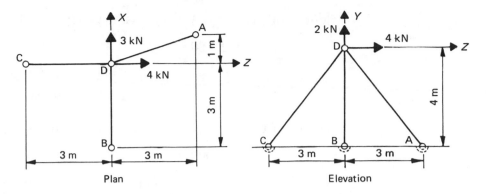

**Figure 2.36**

Similarly,
($\Sigma Y = 0$)

$$-4.0t_{AD} - 4.0t_{BD} - 4.0t_{CD} + 2.0 = 0 \tag{2}$$

($\Sigma Z = 0$)

$$+3.0t_{AD} - 3.0t_{CD} + 4.0 = 0 \tag{3}$$

(2) Determination of tension coefficients
Substitute values of $t_{BD}$ and $t_{CD}$ from Equations (1) and (3), respectively, into Equation (2), to give

$$-4.0t_{AD} - 4.0\frac{(1.0t_{AD} + 3.0)}{3.0} - 4.0\frac{(3.0t_{AD} + 4.0)}{3.0} + 2.0 = 0$$

$$-9.333t_{AD} - 7.333 = 0$$

$$\therefore \underline{t_{AD} = -0.786}$$

Substitute value of $t_{AD}$ into Equation (1):

$$+1.0(-0.786) - 3.0t_{BD} + 3.0 = 0$$

$$\therefore \underline{t_{BD} = +0.738}$$

Substitute value of $t_{AD}$ into Equation (3):

$$+3.0(-0.786) - 3.0t_{CD} + 4.0 = 0$$

$$\therefore \underline{t_{CD} = +0.547}$$

(3) Determination of forces in members
By inspection of Figure 2.36:
length of AD $= (x_{AD}{}^2 + y_{AD}{}^2 + z_{AD}{}^2)^{1/2}$
$= \{(+1.0^2) + (-4.0^2) + (+3.0^2)\}^{1/2}$
$= 5.099$ m

$$\therefore F_{AD} = t_{AD} \times \text{length of AD} = -0.786 \times 5.099$$

$$\therefore \underline{F_{AD} = -4.01 \text{ kN}}$$

length of BD = $\{(-3.0^2) + (-4.0^2) + (0.0^2)\}^{1/2}$

= 5.000 m

$$\therefore F_{BD} = +0.738 \times 5.000$$

$$\therefore F_{BD} = +3.69 \text{ kN}$$

length of CD = $\{(0.0^2) + (-4.0^2) + (-3.0^2)\}^{1/2}$

= 5.000 m

$$\therefore F_{CD} = +0.547 \times 5.000$$

$$\therefore F_{CD} = +2.74 \text{ kN}$$

*Summary*

| Member | Force | Type |
|--------|-------|------|
| AD | 4.01 kN | compressive |
| BD | 3.69 kN | tensile |
| CD | 2.74 kN | tensile |

## Example 2.10

Construct the force diagram for the frame shown in Figure 2.37 and hence determine the magnitude and the sense of the forces in all members.

(Coventry University)

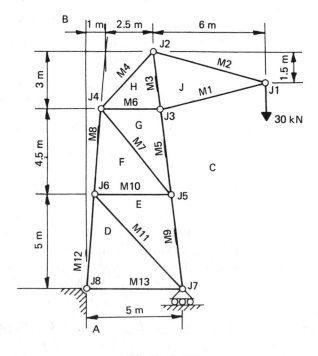

**Figure 2.37**

60

The solution to this problem would be completely graphical, with no calculations or written explanation expected by the examiner. The following explanation is given here to enable the reader to interpret Figures 2.38 and 2.39. In order that reference may be made to certain joints and members during the solving of this problem, each joint and each member has been given an identifying number, as shown in Figure 2.37. The joints and members were not annotated in this way in the original question.

### Solution 2.10

Since there are only two unknown forces at joint J1, the point of application of the load, the force diagram can be commenced at that joint without the necessity of first calculating the reactions.

Using Bow's notation and working clockwise round joint J1, force BC denotes the load of 30 kN, force CJ denotes the force in member M1 and force JB denotes the force in member M2. The triangle of forces at joint J1 is constructed by drawing BC parallel to the load force BC and to some convenient scale representing the magnitude of the load (30 kN). A line is drawn from C parallel to CJ (member M1) and from B parallel to JB (member M2) to intersect at J. The triangle BCJ is the triangle of forces for the three forces acting at joint J1. The length of the vector *JB* can be measured to give the magnitude of force JB and the length of the vector *JC* can be measured to give the magnitude of force JC. The direction of action of the force CJ at joint J1 is given by the direction of the vector *CJ* (acts upwards and to the right). The direction of action of the force JB at joint J1 is given by the vector *JB* (upwards and to the left). These directions are noted by inserting arrowheads on Figure 2.39, the arrowheads being inserted as soon as the corresponding part of the force diagram has been completed. Thus, as soon as the triangle of forces BCJ is drawn, the direction arrows for forces CJ and JB are added to a diagram of the frame. Note that the force in a member acts in opposite directions at the two ends; consequently, a line in the force diagram may be read as two vectors. For example, line JB may be read as vector *JB* (upwards and to the left) or as vector *BJ* (downwards and to the right). In order to select the correct vector for a particular end of a member, it is necessary to identify the member in Figure 2.37 by reading *clockwise* round the joint at that end of the member. Thus, reading clockwise round joint J1, the force in member M2 at joint J1 is given by vector *JB* (upwards and to the left). The direction of force in the other end of member M2 is given by the vector identified by reading *clockwise* round joint J2—that is, by vector *BJ*.

Proceeding with the force diagram, a force triangle is now drawn for joint J2 (only two unknown forces, since the force in member M2 is now determined). Working *clockwise* (it is essential to be consistent) round joint J2, the force in member M2 is denoted by vector *BJ* and this is already in the force diagram as line BJ. From J a line is drawn parallel to member M3 and from B a line is drawn parallel to member M4 to intersect at H. Vector *JH* then gives the magnitude and direction of the force in member M3 at joint J2 and vector *HB* gives the magnitude and direction of the force in member M4 at joint J2.

Proceed to joint J3 (now only two unknown forces at this joint). A line is drawn from C parallel to member M5 and a line is drawn from H parallel to

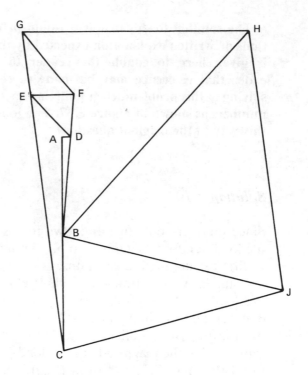

**Figure 2.38**

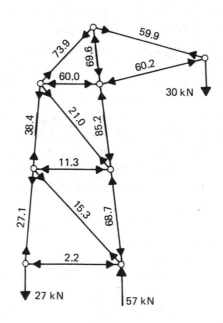

**Figure 2.39**

member M6 to intersect at G and thus complete the four-sided polygon of forces (JCGH) for the four forces acting at joint J3.

Then:

At joint J4, draw from G parallel to member M7 and draw from B parallel to member M8 to intersect at F;

62

At joint J5, draw from C parallel to member M9 and draw from F parallel to member M10 to intersect at E;
At joint J6, draw from E parallel to member M11 and draw from B parallel to member M12 to intersect at D;
At joint J7, draw from C a vertical line (i.e. parallel to the reaction at joint J7) and draw from D parallel to member M13 to intersect at A.

The force diagram is then complete, as in Figure 2.38, and the values of all member forces (and external reactions), can be scaled off and recorded on a diagram of the frame, as in Figure 2.39. The direction of action of the forces can be determined directly from the force diagram. For example, vector **JB** represents a force of magnitude 59.9 kN acting upwards and to the left at the end of member M2 nearest to joint J1, while vector **BJ** represents a force of the same magnitude acting downwards and to the right at the end of member M2 nearest to joint J2. The member is in tension and is a tie.

# 2.5  **Problems**

**2.1**  The pin-jointed truss shown in Figure P2.1 carries vertical loads at C and F of 100 kN and 60 kN, respectively. Determine the forces in all the members, indicating whether they are tensile or compressive.

(Salford University)

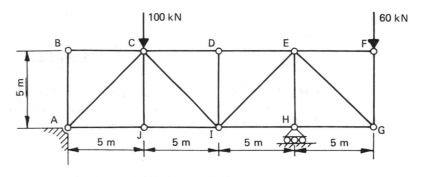

**Figure P2.1**

**2.2**  A series of pin-jointed frameworks, each loaded as shown in Figure P2.2, forms the roof structure of a loading bay. The left-hand support, A, is pinned to the top

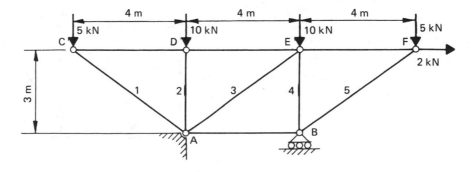

**Figure P2.2**

of a column and the right-hand support, B, rests on rollers on the top of another column. Determine:

(a) the components of the reactions at A and B;
(b) the magnitude and the sense of the force in each of the members marked 1–5 on the figure.

(Coventry University)

2.3 (a) The pin-jointed frame shown in Figure P2.3 is used as part of a canopy and carries loads at B and C as shown. Determine:

(i) the force in every member of the frame, indicating clearly which is tensile and which is compressive;
(ii) the values of the support reactions, indicating clearly their direction.

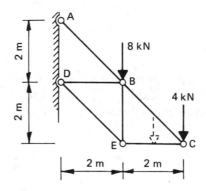

**Figure P2.3**

(b) A further 1 kN load is applied (shown by a broken line) at the centre of member EC. Discuss, without further calculations, why this would normally be bad practice and what the probable result would be.

(University of Portsmouth)

2.4 Determine the force in each member of the pin-jointed truss due to the load system shown in Figure P2.4. Indicate in each case whether the force is tensile or compressive.

(University of Westminster)

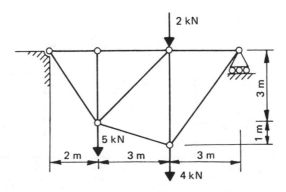

**Figure P2.4**

**2.5** The pin-jointed frame shown in Figure P2.5 is loaded as shown. Determine the magnitude and the nature of the forces in members CD, DJ and CJ.

(University of Hertfordshire)

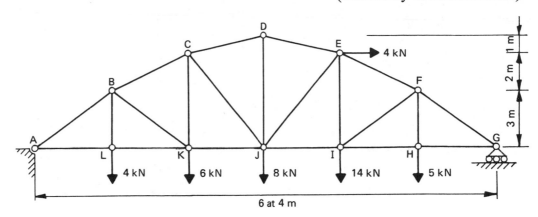

**Figure P2.5**

**2.6** The pin-jointed framework loaded as shown in Figure P2.6 forms a three-pinned frame. Calculate the forces in all the members to the right of the pivot pin, C.

(University of Hertfordshire)

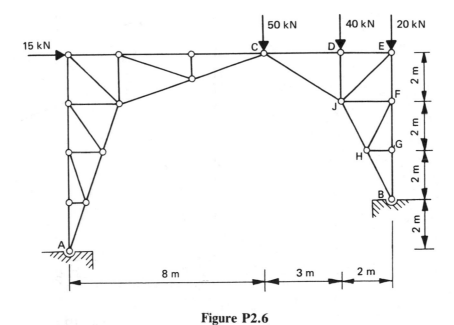

**Figure P2.6**

**2.7** A ball-jointed structure, shown in elevation and plan in Figure P2.7, consists of three steel bars connected at point D. The three support points A, B and C lie in a vertical plane, with A and B at the same level. A force of 10 kN acts vertically down through D. Determine the forces in the bars.

(Cambridge University)

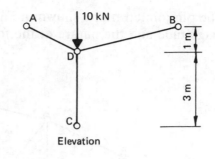

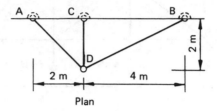

Elevation

Plan

**Figure P2.7**

**2.8**  Construct the force diagram for the framework shown in Figure P2.8 and hence determine the magnitude and sense of the force in each member.

(Coventry University)

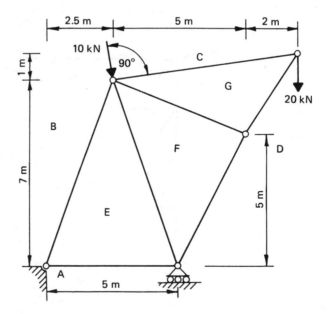

**Figure P2.8**

## 2.6  Answers to Problems

**2.1**

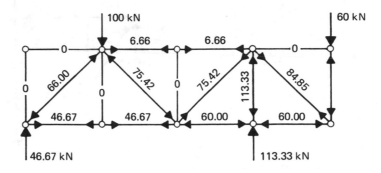

**Figure P2.9**

**2.2**

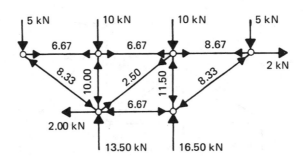

**Figure P2.10**

**2.3**

(a)

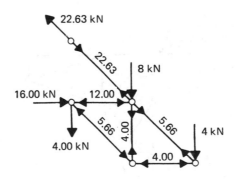

**Figure P2.11**

(b) The analysis of pin-jointed frames assumes that all loads are applied at the joints such that all members are subjected only to axial loads. Thus, member EC will have been designed to resist axial forces only and, consequently, could fail under the action of the bending moment induced by the additional load.

**2.4**

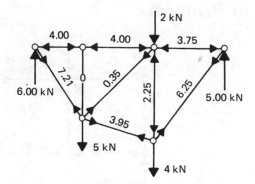

**Figure P2.12**

**2.5**

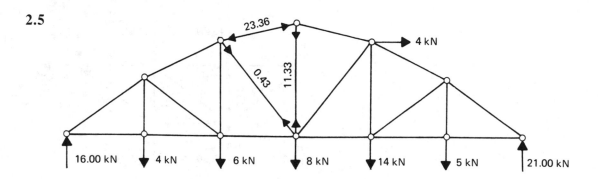

**Figure P2.13**

**2.6**

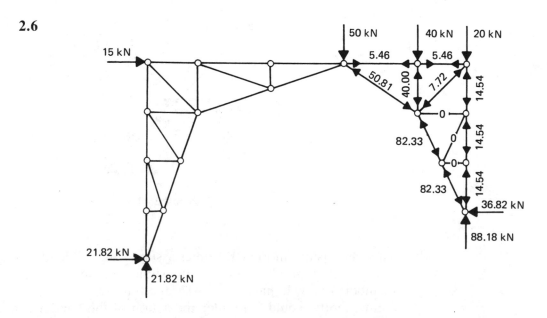

**Figure P2.14**

68

**2.7**

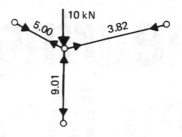

**Figure P2.15**

**2.8**

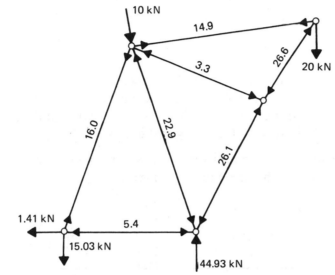

**Figure P2.16**

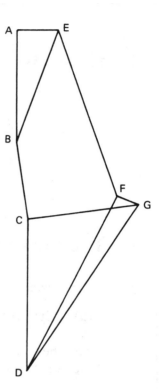

**Figure P2.17**

# 3 Shearing Forces and Bending Moments

## 3.1 Contents

Determination of the values of shearing force and bending moment in beams and in the members of simple portals ● Construction of shearing force and bending moment diagrams resulting from point loads, uniformly distributed loads, non-uniformly distributed loads and couples.

Determination of points of contraflexure, and of the position and magnitude of points of maximum bending moment.

## 3.2 The Fact Sheet

### (a) General Definition of Shearing Force

The shearing force at any section in a horizontal beam is the algebraic sum of all the vertical forces to the left, or to the right, of that section.

### (b) General Definition of Bending Moment

The bending moment at any section in a beam is the algebraic sum of the moments about that section of all the forces to the left, or to the right, of that section.

### (c) Shearing Force Diagrams

A shearing force diagram is a graph showing the variation of the shearing force along the length of a beam or other structural member.

Shearing force diagrams resulting from the action of concentrated loads on horizontal beams consist of horizontal straight lines with vertical jumps at load positions.

Shearing force diagrams resulting from the action of uniformly distributed loads on horizontal beams consist of sloping straight lines.

70

### (d) Bending Moment Diagrams

A bending moment diagram is a graph showing the variation of the bending moment along the length of a beam or other structural member.

Bending moment diagrams resulting from the action of concentrated point loads consist of straight lines with changes of direction at the points of application of the loads.

Bending moment diagrams resulting from the action of uniformly distributed loads consist of parabolic curves.

At the point of application of a couple to a horizontal beam, there will be a vertical step in the bending moment diagram.

The bending moment at an internal pin or a pinned support is zero.

### (e) Points of Contraflexure

The direction of curvature changes at a point of contraflexure—that is, at a point where the bending moment is zero.

### (f) Standard Formulae for Maximum Bending Moment

If a single concentrated vertical load $W$ acts at the mid-point of a simply supported beam of span $L$, then the maximum bending moment is at the point of application of the load and is of magnitude $WL/4$.

If a uniformly distributed load $w$ per unit length acts along the entire length of a simply supported beam of span $L$, then the maximum bending moment is at mid span and is of magnitude $wL^2/8$.

## 3.3  Symbols, Units and Sign Conventions

$H$    = horizontal reaction at a support (kN)
$V$    = vertical reaction at a support (kN)
B.M. = bending moment (kN m)
S.F.  = shearing force (kN)

The shearing force at a section in a beam is taken to be positive if the resultant vertical loading on the part of the beam to the left of that section tends to move that part of the beam vertically upwards.

The bending moment in a beam is taken to be positive if it causes the beam to sag.

Bending moment diagrams are plotted such that positive bending moments are on the tensile side of the beam or other structural member.

## 3.4 Worked Examples

### Example 3.1

The beam ABCDE shown in Figure 3.1 carries a uniformly distributed load of 6 kN/m over BC, plus a point load of 20 kN at D. Draw the shearing force and bending moment diagrams for the beam, indicating all significant values.

(University of Hertfordshire)

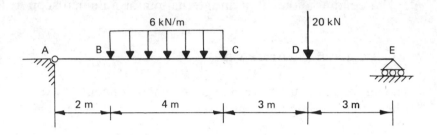

**Figure 3.1**

### Solution 3.1

(1) To determine the reactions
Taking moments about A:
$(\Sigma M_A = 0)$

$$+(6 \times 4 \times 4) + (20 \times 9) - (V_E \times 12) = 0$$

$$\therefore V_E = +23.00 \text{ kN}$$

Resolving vertically:
$(\Sigma V = 0)$

$$V_A + V_E - (6 \times 4) - 20 = 0$$

$$\therefore V_A = +21.00 \text{ kN}$$

Check by taking moments about E:
$(\Sigma M_E = 0)$

$$+(21 \times 12) - (6 \times 4 \times 8) - (20 \times 3) = 0$$

$$\therefore \text{ correct}$$

(2) Shearing force diagram (working from *left to right* and summing the forces to the left of the point under consideration)

$$\text{Just to the right of A, S.F.} = V_A = +21.00 \text{ kN}$$
$$\text{To the left of B, S.F.} = V_A = +21.00 \text{ kN}$$
$$\text{At C, S.F.} = V_A - (6 \times 4) = +21.00 - 24.00 = -3.00 \text{ kN}$$

**Between B and C the graph will be linear with a slope of −6 kN/m.**

Just to the left of D, S.F. $= V_A - (6 \times 4) = +21.00 - 24.00 = -3.00$ kN
To the right of D, S.F. $= -3.00 - 20.00 = -23.00$ kN
To the left of E, S.F. $= -3.00 - 20.00 = -23.00$ kN

**The shearing force diagram is shown in Figure 3.2.**

(3)   Bending moment diagram (working from left to right and summing the moments of all forces to the left of the section under consideration)

**It will be necessary to quote the value of the maximum bending moment, which will occur at the point of zero shearing force. The position of the point of zero shearing force may be calculated from the geometry of the shearing force diagram (see Figure 3.2), using similar triangles.**

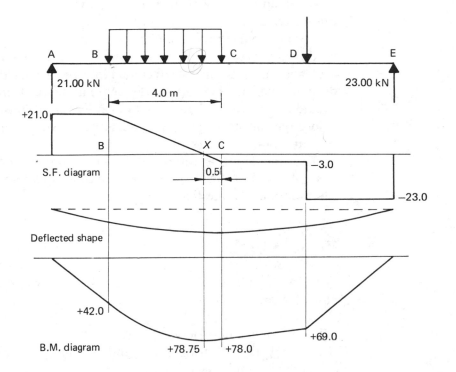

**Figure 3.2**

If $X$ is the point of zero shearing force,

$$\frac{CX}{CB} = \frac{3}{3 + 21}$$

i.e.

$$\frac{CX}{4} = \frac{3}{3 + 21}$$

$$\therefore CX = 0.5 \text{ m}$$

B.M. at A $= 0 = 0.00$ kNm
B.M. at B $= +(21.00 \times 2) = +42.00$ kNm

73

The effect of the upward acting reaction at A is to bend the beam in a curved shape sagging downwards. The bending moment at B is thus positive.

$$B.M. \text{ at } C = +(21.00 \times 6) - (6.00 \times 4 \times 2) = +78.00 \text{ kNm}$$
$$B.M. \text{ at } D = +(21.00 \times 9) - (6.00 \times 4 \times 5) = +69.00 \text{ kNm}$$
$$B.M. \text{ at } E = 0.00 \text{ kNm}$$

Between A and B, between C and D, and between D and E the graph will be linear, because there are no loads on the beam between those points. Between B and C the graph will be parabolic, with a maximum value at $X$.

$$B.M. \text{ at } X = +(21.00 \times 5.5) - (6.00 \times 3.5 \times 1.75) = \underline{+78.75 \text{ kNm}}$$

The bending moment diagram is shown in Figure 3.2. It is recommended that the diagrams be drawn in vertical projection as shown, so that corresponding values of shearing force and bending moment at any section of the beam are readily apparent. It is also recommended that the deflected shape of the beam be sketched, because an appreciation of the deflected shape is of assistance in deriving the correct shape of the bending moment diagram. The deflected shape should be drawn first and used to identify the regions (top or bottom) of the beam where tension develops, thus indicating on which side of the beam to draw the bending moment diagram.

Diagrams showing the deflected shapes are included for the majority of the examples and problems in this text, even though they are not always asked for in the examination questions.

## Example 3.2

A beam ABC is supported at A and B as shown in Figure 3.3. For the loading case shown:

(i)   calculate the reactions at A and B;
(ii)  draw the shearing force and bending moment diagrams, indicating all important values;
(iii) determine the position and magnitude of the maximum sagging bending moment in span AB.

(University of Portsmouth)

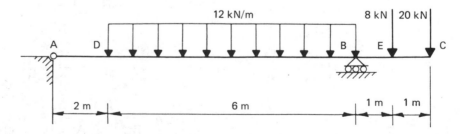

**Figure 3.3**

74

*Solution 3.2*

(1) To determine the support reactions
Taking moments about A:
$(\Sigma M_A = 0)$
$$+(12.0 \times 6 \times 5) - (V_B \times 8) + (8.0 \times 9) + (20.0 \times 10) = 0$$
$$\therefore V_B = +79.00 \text{ kN}$$

Resolving vertically:
$(\Sigma V = 0)$
$$V_A - (12.0 \times 6) + V_B - 8.0 - 20.0 = 0$$
$$\therefore V_A = +21.00 \text{ kN}$$

Check by taking moments about B:
$(\Sigma M_B = 0)$
$$+(21.0 \times 8) - (12.0 \times 6 \times 3) + (8.0 \times 1) + (20.0 \times 2) = 0 \text{ (correct)}$$

(2) Shearing force diagram

$$\text{Just to the right of A, S.F.} = V_A = +21.00 \text{ kN}$$
$$\text{To the left of D, S.F.} = V_A = +21.00 \text{ kN}$$
$$\text{To the left of B, S.F.} = +21.00 - (12.00 \times 6) = -51.00 \text{ kN}$$
$$\text{To the right of B, S.F.} = +21.00 - (12.00 \times 6) + 79.00 = +28.00 \text{ kN}$$
$$\text{To the left of E, S.F.} = +21.00 - (12.00 \times 6) + 79.00 = +28.00 \text{ kN}$$
$$\text{To the right of E, S.F.} = +21.00 - (12.00 \times 6) + 79.00 - 8.00 = +20.00 \text{ kN}$$
$$\text{To the left of C, S.F.} = +21.00 - (12.00 \times 6) + 79.00 - 8.00 = +20.00 \text{ kN}$$

(3) Bending moment diagram

**Considering moments of forces to the left:**

$$\text{B.M. at A} = 0.00 \text{ kNm}$$
$$\text{B.M. at D} = +(21.00 \times 2) = +42.00 \text{ kNm}$$
$$\text{B.M. at B} = +(21.00 \times 8) - (12.00 \times 6 \times 3) = -48.00 \text{ kNm}$$

**Considering moments of forces to the right:**

$$\text{B.M. at E} = -(20.00 \times 1) = -20.00 \text{ kNm}$$
$$\text{B.M. at C} = 0.00 \text{ kNm}$$

**Note that the calculation of the bending moment at E has been simplified by considering forces to the right of E rather than those to the left.**

(4) Maximum sagging bending moment
If X is the point of zero shearing force (see Figure 3.4), then, by similar triangles,

$$\frac{DX}{DB} = \frac{DX}{6} = \frac{21.00}{21.00 + 51.00}$$
$$\therefore DX = 1.75 \text{ m}$$

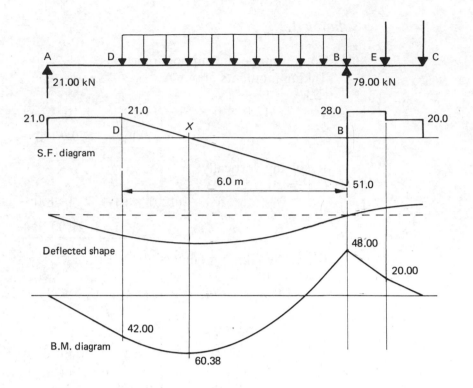

**Figure 3.4**

Maximum B.M. is at $X$, i.e. <u>1.75 m</u> to the right of D, and the value of the B.M. at $X$ = +(21.00 × 3.75) − (12.00 × 1.75 × 1.75/2) = <u>+60.38 kNm</u>

**The shearing force and bending moment diagrams are shown in Figure 3.4.**

## Example 3.3

For the simply supported beam loaded as shown in Figure 3.5:

  (i)  establish equations for the shearing force and for the bending moment at a section $X$–$X$ distance $x$ from A;
  (ii)  determine the position and value of the maximum bending moment in the beam;
  (iii)  sketch the bending moment diagram for the beam.

(Coventry University)

## *Solution 3.3*

(1)  To determine the support reactions

**The load on this beam increases uniformly from zero to 60 kN/m over a length of 2.4 m. The total load will be obtained by multiplying the average intensity of load,**

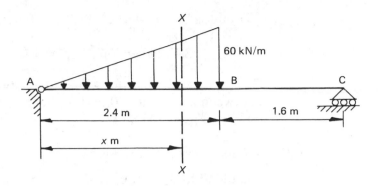

**Figure 3.5**

**30 kN/m, by 2.4 m. This total load (i.e. the resultant of the load system) will act through the centroid of the load distribution diagram (i.e. at $\frac{2}{3}$ of 2.4 m from A).**

Total load = $30 \times 2.4 = 72.00$ kN acting at 1.6 m to the right of A.
Taking moments about A:
$(\Sigma M_A = 0)$

$$+(72.00 \times 1.6) - (V_C \times 4) = 0$$

$$\therefore V_C = +28.80 \text{ kN}$$

Resolving vertically:
$(\Sigma V = 0)$

$$+V_A - 72.00 + V_C = 0$$
$$V_A - 72.00 + 28.80 = 0$$

$$\therefore V_A = +43.20 \text{ kN}$$

(2)   Shearing force at section $X$–$X$

**It is necessary to calculate the value of the total load to the left of section $X$–$X$. The intensity of load at section $X$–$X$ may be derived by the use of similar triangles in Figure 3.6.**

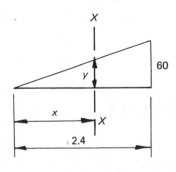

**Figure 3.6**

Total load to the left of $X$–$X$ = $\frac{1}{2}(y \times x)$
where $y/60 = x/2.4$.

77

$\therefore$ Total load to the left of $X–X = \frac{1}{2}(60x/2.4) \times x$
$$= 12.5x^2$$
$\therefore$ S.F. at section $X–X = +V_A - 12.5x^2 = +(43.20 - 12.5x^2)$ kN

This expression is true for values of $x$ between zero and 2.4 m. For values of $x$ between 2.4 m and 4.0 m, the shearing force at $X–X$ will be constant and equal to $V_C = -28.80$ kN.

(3) Bending moment at section $X–X$

Total load to the left of $X–X = 12.5x^2$ acting at $2x/3$ from A.
$\therefore$ B.M. at $X–X = +(V_A \times x) - (\text{load} \times (x - 2x/3))$
$$= +43.20x - (12.5x^2) \times x/3$$
$$= +43.20x - 4.17x^3 \text{ kNm} \qquad (1)$$

This expression is valid for values of $x$ between zero and 2.4 m.
For values of $x$ between 2.4 m and 4.0 m,

$$\text{B.M. at } X–X = +V_C \times (4.0 - x) = 28.80 \times (4.0 - x)$$
$$= +115.2 - 28.8x \text{ kNm}$$

(4) Position and magnitude of the maximum bending moment
The maximum bending moment occurs at the section where the shearing force is zero. From the shearing force diagram (see Figure 3.7), it is clear that this will occur somewhere between A and B. The shearing force at section $X–X = +43.20 - 12.5x^2$.

$$\therefore \text{S.F.} = 0 \text{ when } x = \left(\frac{43.20}{12.5}\right)^{1/2} = 1.86 \text{ m}$$

Therefore, maximum B.M. occurs at 1.86 m from A.
From Equation (1),

$$\text{maximum B.M.} = +(43.20 \times 1.86) - (4.17 \times 1.86^3) = +53.52 \text{ kNm}$$

**The B.M. diagram together with the S.F. and deflected shape diagrams are shown in Figure 3.7.**

**Example 3.4**

(a) The simply supported beam ABC shown in Figure 3.8(a) is subjected to the loading shown. Draw the shearing force and bending moment diagrams for the beam.

(b) A similar beam ABCD (Figure 3.8b) has an overhanging section CD loaded as shown. Using superposition or otherwise, draw the shearing force and bending moment diagrams for ABCD and sketch the deflected shape.

(Salford University)

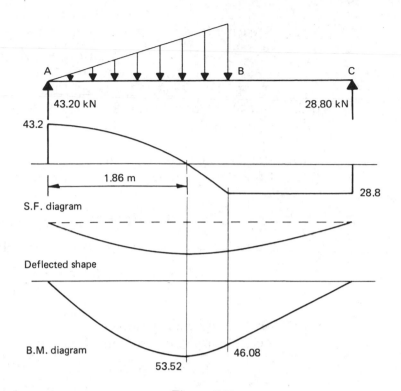

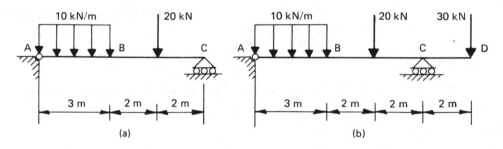

**Figure 3.7**

**Figure 3.8**

*Solution 3.4*

(a)

(1)   To determine the support reactions
      Taking moments about A:
      $(\Sigma M_A = 0)$

$$+(10.0 \times 3 \times \tfrac{3}{2}) + (20.0 \times 5) - (V_C \times 7) = 0$$

$$\therefore \underline{V_C = +20.71 \text{ kN}}$$

Resolving vertically:
$(\Sigma V = 0)$

$$+V_A - (10.0 \times 3) - 20.0 + V_C = 0$$

$$\therefore \underline{V_A = +29.29 \text{ kN}}$$

(2) Shearing force diagram

$$\text{Just to the right of A, S.F.} = V_A = +29.29 \text{ kN}$$
$$\text{To the left of B, S.F.} = +29.29 - (10.0 \times 3) = -0.71 \text{ kN}$$
$$\text{To the left of the 20 kN load, S.F.} = +29.29 - (10.0 \times 3) = -0.71 \text{ kN}$$
$$\text{To the right of the 20 kN load, S.F.} = +29.29 - (10.0 \times 3) - 20.0 = -20.71 \text{ kN}$$
$$\text{Just to the left of C, S.F.} = +29.29 - (10.0 \times 3) - 20.0 = -20.71 \text{ kN}$$

(3) Bending moment diagram
Considering moments of forces to the left:

$$\text{B.M. at A} = 0.00 \text{ kNm}$$
$$\text{B.M. at B} = +(29.29 \times 3) - (10.0 \times 3 \times \tfrac{3}{2}) = +42.87 \text{ kNm}$$

Considering moments of forces to the right:

$$\text{B.M. at 20 kN load} = +20.71 \times 2 = +41.42 \text{ kNm}$$
$$\text{B.M. at C} = 0.00 \text{ kNm}$$

(4) Maximum bending moment
Zero shearing force occurs at $x$ from A (see Figure 3.9), where

$$\frac{x}{3} = \frac{29.29}{29.29 + 0.71}$$

i.e. $x = 2.93$ m
Therefore, the maximum B.M. occurs at 2.93 m from A.

$$\text{Value of maximum B.M} = +(29.29 \times 2.93) - (10.0 \times 2.93 \times 2.93/2)$$
$$= +42.89 \text{ kNm}$$

**The shearing force and bending moment diagrams are shown in Figure 3.9.**

(b)
**The principle of superposition will be used to solve part (b). The shearing force and bending moments in the beam due to all the loads acting together may be determined by adding the values due to the additional 30.0 kN load acting alone to the corresponding values obtained in part (a).**

(1) Considering the 30.0 kN load acting alone (i.e. assuming that the only load on ABCD is the 30 kN load)
Taking moments about A:
$(\Sigma M_A = 0)$

$$+(30.0 \times 9) - (V_C \times 7) = 0$$
$$\therefore V_C = +38.57 \text{ kN}$$

Resolving vertically:
$(\Sigma V = 0)$

$$+V_A + 38.57 - 30.00 = 0$$
$$\therefore V_A = -8.57 \text{ kN}$$

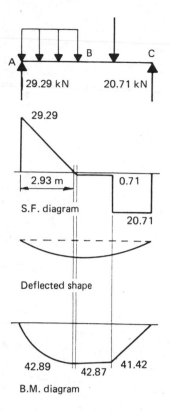

**Figure 3.9**

(2)  Shearing force diagram

Just to the right of A, S.F. = $V_A$ = −8.57 kN

To the left of C, S.F. = $V_A$ = −8.57 kN

To the right of C, S.F. = −8.57 + 38.57 = +30.00 kN

To the left of D, S.F.= −8.57 + 38.57 = +30.00 kN

(3)  Bending moment diagram

B.M. at A = 0.00 kNm

Considering moments of forces to the left:

B.M. at C = +(−8.57 × 7) = −60.00 kNm

Alternatively, considering moments of forces to the right,

B.M. at C = −30.00 × 2 = −60.00 kNm

B.M. at D = 0.00 kNm

**The shearing force and bending moment diagrams for the beam with the 30.00 kN load acting only are shown in Figure 3.10.**

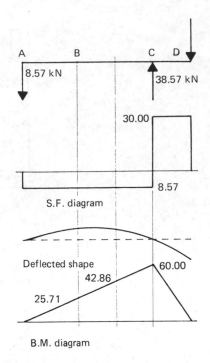

**Figure 3.10**

(4) Shearing force and bending moment diagrams for the complete load system.

**The shearing force and bending moment diagrams for beam ABCD with the complete load system are shown in Figure 3.11, which is obtained by adding together the ordinates of the diagrams in Figures 3.9 and 3.10.**

If the shearing force is zero at distance $x$ from A, then, from the shearing force diagram in Figure 3.11,

$$\frac{x}{3} = \frac{20.72}{20.72 + 9.28}$$

$$\therefore x = 2.07 \text{ m}$$

Therefore, the maximum B.M. is at 2.07 m from A.

Value of maximum B.M. $= +(20.72 \times 2.07) - (10.00 \times 2.07 \times 2.07/2)$
$= +21.47$ kNm

## Example 3.5

The simply supported beam in Figure 3.12 carries a vertical load of 20 kN at 4 m from A. Welded to the beam at B are two vertical arms each carrying a horizontal load of 15 kN as shown. Draw the shearing force and bending moment diagrams for the beam, marking all important values, and sketch the deformed shape of the beam.

(Manchester University)

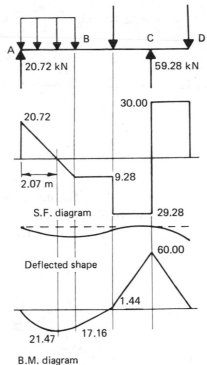

**Figure 3.11**

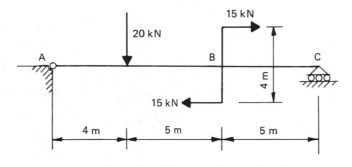

**Figure 3.12**

The forces on the two vertical arms constitute a couple applying a clockwise moment of value **15.00 × 4 = +60.00 kNm to the beam at point B. Thus, there will be a vertical jump of 60.00 kNm in the bending moment diagram at point B.**

*Solution 3.5*

(1)  To determine the support reactions
Taking moments about A:
$(\Sigma M_A = 0)$

$$+(20.0 \times 4) + (60.00) - (V_C \times 14) = 0$$

$$\therefore V_C = +10.00 \text{ kN}$$

Resolving vertically:
($\Sigma V = 0$)

$$+V_A - 20.00 + 10.00 = 0$$

$$\therefore V_A = +10.00 \text{ kN}$$

(2)  Shearing force diagram

Just to the right of A, S.F. $= V_A = +10.00$ kN
To the left of the 20 kN load, S.F. $= V_A = +10.00$ kN
To the right of the 20 kN load, S.F. $= +10.00 - 20.00 = -10.00$ kN
To the left of C, S.F. $= +10.00 - 20.00 = -10.00$ kN

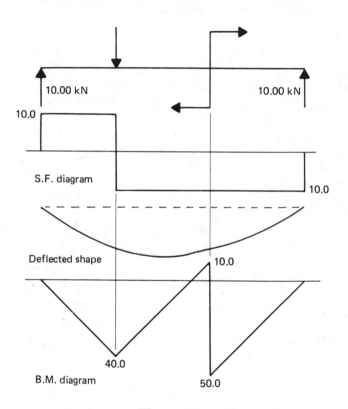

**Figure 3.13**

(3)  Bending moment diagram
Considering moments of forces to the left:

B.M. at A $= 0.00$ kNm
B.M. at 20 kN load $= +(V_A \times 4) = +40.00$ kNm
B.M. just to the left of B $= +(V_A \times 9) - (20.00 \times 5) = -10.00$ kNm
B.M. just to the right of B $= +(V_A \times 9) - (20.00 \times 5) + 60.00 = +50.00$ kNm
B.M. at C $= 0.00$ kNm

84

**Example 3.6**

The diagram of Figure 3.14 shows a cranked beam ABCDEFG which is supported on a knife edge at A, and a knife-edge and roller system at F. The upstand BH is rigidly connected at B.

Make dimensioned sketches which show the thrusts, shearing forces and bending moments which will arise in the structure due to the application of the loads shown. Clearly indicate the sign convention used.

(Sheffield University)

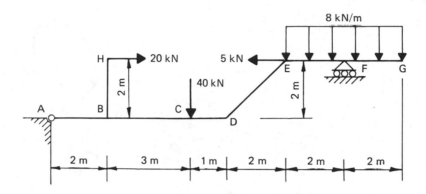

**Figure 3.14**

*Solution 3.6*

(1) To determine the support reactions
Taking moments about A:
$(\Sigma M_A = 0)$

$$+(20 \times 2) + (40 \times 5) - (5 \times 2) + (8 \times 4 \times 10) - (V_F \times 10) = 0$$

$$\therefore V_F = +55.00 \text{ kN}$$

Resolving vertically:
$(\Sigma V = 0)$

$$+V_A - 40 - (8 \times 4) + 55.00 = 0$$

$$\therefore V_A = +17.00 \text{ kN}$$

Resolving horizontally:
$(\Sigma H = 0)$

$$H_A + 20 - 5 = 0$$

$$\therefore H_A = -15.00 \text{ kN}$$

**Note that it is assumed that the knife-edge support at A can provide a horizontal reaction due to friction between the beam and the support.**

85

(2)   Shearing force diagram

$$\text{Just to the right of A, S.F.} = V_A = +17.00 \text{ kN}$$
$$\text{To the left of C, S.F.} = V_A = +17.00 \text{ kN}$$
$$\text{To the right of C, S.F.} = +17.00 - 40.00 = -23.00 \text{ kN}$$

**For the length DE the shearing forces will be the net forces acting at right angles to the beam. Just to the right of D there is a vertical force of 23.00 kN and a horizontal force of 5.00 kN (due to the longitudinal thrust in BCD). Thus, the shearing force in DE at D is the sum of the components of these two forces resolved in a direction normal to the beam DE. Refer to Figure 3.15, which shows the free body diagram for DE.**

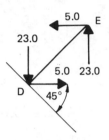

**Figure 3.15**

$$\text{S.F. in DE just to the right of D} = -23.00 \cos 45° - 5.00 \cos 45°$$
$$= -19.80 \text{ kN}$$

**This value of shearing force will be constant along the length DE, since no external loads act on this length of the beam. Just to the right of E the vertical shearing force will be the same as the shearing force to the left of D.**

$$\text{To the right of E, S.F.} = +17.00 - 40.00 = -23.00 \text{ kN}$$
$$\text{To the left of F, S.F.} = +17.00 - 40.00 - (8 \times 2) = -39.00 \text{ kN}$$
$$\text{To the right of F, S.F.} = +17.00 - 40.00 - (8 \times 2) + 55.00 = +16.00 \text{ kN}$$
$$\text{To the left of G, S.F.} = +17.00 - 40.00 - (8 \times 4) + 55.00 = 0.00 \text{ kN}$$

(3)   Bending moment diagram

$$\text{B.M. at A} = 0.00 \text{ kNm}$$
$$\text{B.M. just to the left of B} = +(V_A \times 2) = +34.00 \text{ kNm}$$
$$\text{B.M. just to the right of B} = +34.00 + (20 \times 2) = +74.00 \text{ kNm}$$

**Note that the upstand crank BH applies a clockwise moment of value 20 × 2 kNm to the beam at B.**
$$\text{B.M. at C} = +(V_A \times 5) + (20 \times 2) = +125.00 \text{ kNm}$$
$$\text{B.M. at D} = +(V_A \times 6) + (20 \times 2) - (40 \times 1) = +102.00 \text{ kNm}$$

**Then, working from the right-hand end of the beam,**

$$\text{B.M. at G} = 0.00 \text{ kNm}$$
$$\text{B.M. at F} = -(8 \times 2 \times 2/2) = -16.00 \text{ kNm}$$
$$\text{B.M. at E} = -(8 \times 4 \times 4/2) + (V_F \times 2) = +46.00 \text{ kNm}$$

(4)  Thrusts

Longitudinal thrust in AB = $H_A$ = +15.00 kN (i.e. tensile)
Longitudinal thrust in BCD = $H_A$ − 20 = −5.00 kN (i.e. compressive)

**To determine the longitudinal thrust in DE refer to Figure 3.15 and resolve along the member DE:**

Longitudinal thrust in DE = 23 sin45° − 5 sin45° = +12.73 kN
(i.e. tensile)

The thrusts are marked on the appropriate lengths of the structure in the shearing force diagram in Figure 3.16. A compressive thrust is shown as negative, a tensile thrust as positive.

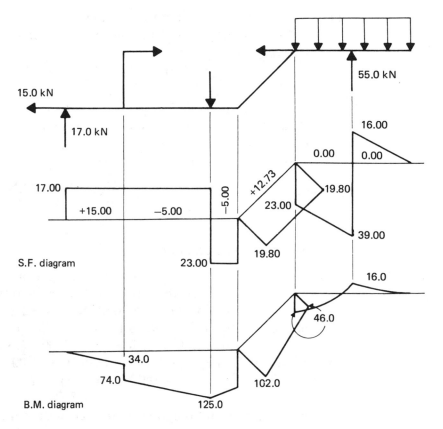

**Figure 3.16**

## Example 3.7

Figure 3.17 shows a beam system which is built in at A, simply supported at C and articulated at B. Construct the shearing force and bending moment diagrams which result from the loading shown.

(University of Portsmouth)

87

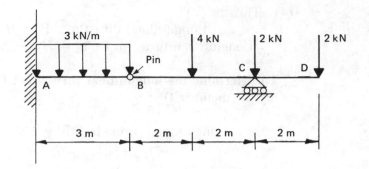

**Figure 3.17**

*Solution 3.7*

(1) To determine the support reactions
   Taking moments about the pin B of the forces to the right of B:
   ($\Sigma M_B = 0$)

   $$+(4 \times 2) + (2 \times 4) + (2 \times 6) - (V_C \times 4) = 0$$

   $$\therefore V_C = +7.00 \text{ kN}$$

   Resolving vertically for the whole structure:
   ($\Sigma V = 0$)

   $$+V_A - (3 \times 3) - 4 - 2 - 2 + V_C = 0$$

   $$\therefore V_A = +10.00 \text{ kN}$$

(2) Shearing force diagram

   $$\text{Just to the right of A, S.F.} = V_A = +10.00 \text{ kN}$$
   $$\text{To the left of B, S.F.} = +10.00 - (3 \times 3) = +1.00 \text{ kN}$$
   $$\text{To the left of the 4 kN load, S.F.} = +10.00 - (3 \times 3) = +1.00 \text{ kN}$$
   $$\text{To the right of the 4 kN load, S.F.} = +10.00 - (3 \times 3) - 4.00 = -3.00 \text{ kN}$$
   $$\text{To the left of C, S.F.} = +10.00 - (3 \times 3) - 4.00 = -3.00 \text{ kN}$$
   $$\text{To the right of C, S.F.} = +10.00 - (3 \times 3) - 4.00 - 2.00 + V_C = +2.00 \text{ kN}$$
   $$\text{To the left of D, S.F.} = +10.00 - (3 \times 3) - 4.00 - 2.00 + V_C = +2.00 \text{ kN}$$

(3) Bending moment diagram **(note that there is a fixing moment at A)**
   Considering the left-hand part AB **(see the free body diagram for AB in Figure 3.18)** and taking moments about A:

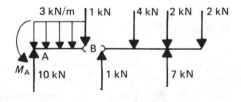

**Figure 3.18**

88

$(\Sigma M_A = 0)$

$$-M_A + (3 \times 3) \times \tfrac{3}{2} + (1 \times 3) = 0$$

$$\therefore M_A = +16.50 \text{ kNm}$$

(i.e. 16.5 kNm anticlockwise).

$$\text{B.M. at B} = 0.00 \text{ kNm}$$
$$\text{B.M. at the 4 kN load} = -M_A + (V_A \times 5) - (3 \times 3) \times 3.5$$
$$= -16.5 + (+10.00 \times 5) - 31.5 = +2.00 \text{ kNm}$$

Considering moments of forces to the right:

$$\text{B.M. at C} = -(2 \times 2) = -4.00 \text{ kNm}$$
$$\text{B.M. at D} = 0.00 \text{ kNm}$$

The shearing force and bending moment diagrams are shown in Figure 3.19.

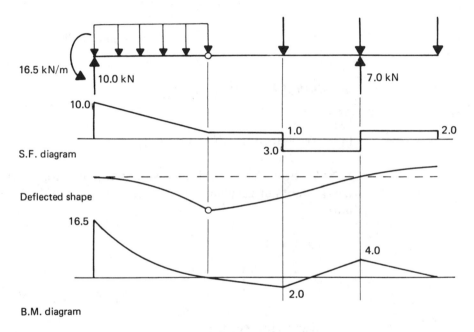

**Figure 3.19**

**If the reader is in any doubt about the shape of the bending moment graph between A and B, the value of bending moment at an intermediate point (e.g. half-way between A and B) should be calculated.**

## Example 3.8

(a) Determine the horizontal and vertical reactions at A and D for the 'bent' shown in Figure 3.20.

(b) Resolve the values of the external forces at A and D and the internal forces at B and C along each member.

(c)  Sketch the shearing force diagram.
(d)  Sketch the bending moment diagram.

(Nottingham University)

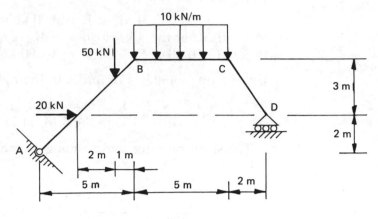

**Figure 3.20**

## Solution 3.8

(a)
(1)  To determine the support reactions

**Note that A is a pinned support providing both horizontal and vertical components of reaction. D is a roller support and will provide a vertical reaction only.**

$$(M_A = 0)$$

$$+(20 \times 2) + (50 \times 4) + (10 \times 5 \times 7.5) - (V_D \times 12) = 0$$

$$\therefore V_D = +51.25 \text{ kN}$$

Resolving vertically:
$$(\Sigma V = 0)$$

$$V_A + V_D - 50 - (10 \times 5) = 0$$

$$\therefore V_A = +48.75 \text{ kN}$$

Resolving horizontally:
$$(\Sigma H = 0)$$

$$H_A + 20 = 0$$

$$\therefore H_A = -20.00 \text{ kN}$$

(b)
(1)  To determine the longitudinal thrusts in the members

**Refer to Figures 3.21 and 3.22 to assess the magnitude and the sense of the thrusts in the members.**

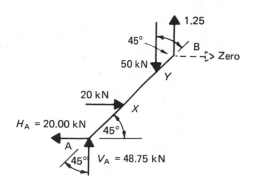

**Figure 3.21**

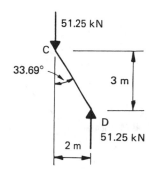

**Figure 3.22**

Resolving along the length of the members:

Between A and $X$, thrust = 48.75 cos45° − 20 cos45°
$\qquad\qquad\qquad\qquad$ = 20.33 kN (compression)
Between $X$ and $Y$, thrust = 48.75 cos45° − 20 cos45° + 20 cos45°
$\qquad\qquad\qquad\qquad$ = 34.47 kN (compression)
Between $Y$ and $B$, thrust = 48.75 cos45° − 20 cos45°
$\qquad\qquad\qquad\qquad$ + 20 cos45° − 50 cos45°
$\qquad\qquad\qquad\qquad$ = −0.88 kN (i.e. 0.88 kN tension)
Between B and C, thrust = 0
Between C and D, (refer to Figure 3.22)
$\qquad\qquad\qquad$ thrust = 51.25 cos33.69° = 42.64 kN (compression)

(c)
(1)  Shearing force diagram

**To construct the complete shearing force diagram, it will be necessary to resolve forces at right angles to the members. Refer to Figures 3.21 and 3.22.**

Just to the right of A (in AB), S.F. = +48.75 sin45° + 20 sin45° = +48.61 kN
To the left of $X$, S.F. = +48.75 sin45° + 20 sin45° = +48.61 kN
To the right of $X$, S.F. = +48.75 sin45° + 20 sin45° − 20 sin45° = +34.47 kN
To the left of $Y$, S.F. = +48.75 sin45° + 20 sin45° − 20 sin45° = +34.47 kN
To the right of $Y$, S.F. = +48.75 sin45° + 20 sin45° − 20 sin45° − 50 sin45° = −0.88 kN

91

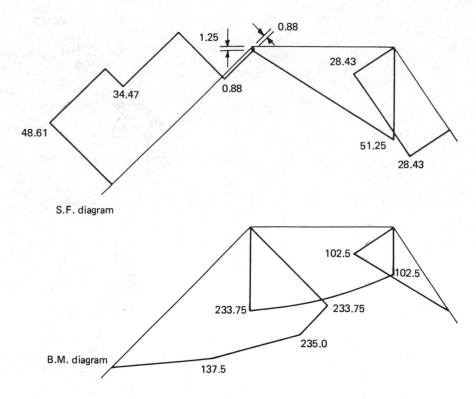

**Figure 3.23**

To the left of B, S.F. $= +48.75 \sin45° + 20 \sin45° - 20 \sin45° - 50 \sin45° = -0.88$ kN
To the right of B (in BC), S.F. $= +48.75 - 50.00 = -1.25$ kN
To the left of C, S.F. $= +48.75 - 50.00 - (10 \times 5) = -51.25$ kN
To the right of C (in CD), S.F. $= -51.25 \sin33.69° = -28.43$ kN
To the left of D, S.F. $= -51.25 \sin33.69° = -28.43$ kN

(d)
(1) Bending moment diagram (**refer to Figures 3.21 and 3.22 for the direction of action of the external forces**)

B.M. at A $= 0.00$ kNm
B.M. at $X = +(V_A \times 2) + (H_A \times 2) = +(48.75 \times 2) + (20 \times 2)$
$= +137.50$ kNm
B.M. at $Y = +(V_A \times 4) + (H_A \times 4) - (20 \times 2)$
$= +(48.75 \times 4) + (20 \times 4) - (20 \times 2) = +235.00$ kNm
B.M. at $B = +(V_A \times 5) + (H_A \times 5) - (20 \times 3) - (50 \times 1)$
$= +(48.75 \times 5) + (20 \times 5) - (20 \times 3) - (50 \times 1)$
$= +233.75$ kNm

**Considering the moment of the force to the right:**

B.M. at C $= +(51.25 \times 2) = +102.50$ kNm

**The shearing force and bending moment diagrams are shown in Figure 3.23.**

# 3.5  Problems

**3.1**  Figure P3.1 shows the details of a loaded beam ACDBE.

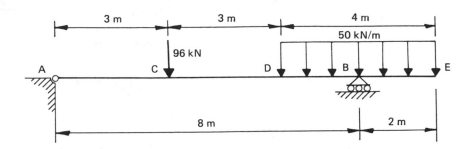

**Figure P3.1**

(a)  Sketch bending moment and shearing force diagrams for this beam, indicating clearly the convention adopted. Show on the diagrams the values of the bending moment and shearing force at points B, C and D.

(b)  State the maximum values of the bending moment and shearing forces and indicate their positions.

(c)  Calculate the position of any point of contraflexure.

(d)  Sketch the deflected form.

(University of Hertfordshire)

**3.2**  Plot the shearing force and bending moment diagrams for the cranked beam shown in Figure P3.2. The bending moment diagram should be plotted on the tension side of the beam. Calculate the maximum sagging bending moment and its distance from the left-hand support, B. Also calculate the distance from B to any point of contraflexure.

(Liverpool University)

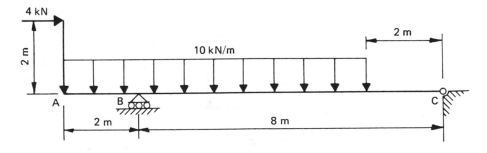

**Figure P3.2**

**3.3**  Construct clear sketches showing bending moment and shearing force diagrams for the beam shown in Figure P3.3. Include salient values on the sketches.
    Determine the position and value of the maximum sagging bending moment in span EC.

(University of Portsmouth)

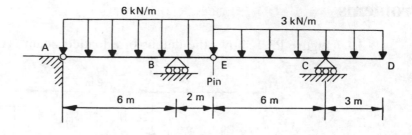

**Figure P3.3**

**3.4** For the cranked beam shown in Figure P3.4 plot the shearing force, thrust and bending moment diagrams.

(Liverpool University)

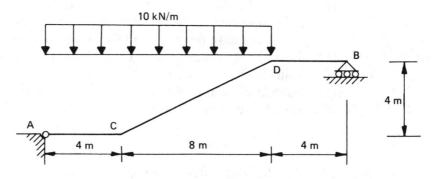

**Figure P3.4**

**3.5** Draw the bending moment diagram for the three-pinned frame shown in Figure P3.5.

(University of Westminster)

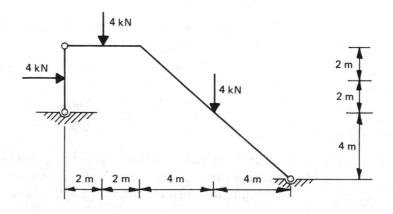

**Figure P3.5**

## 3.6 Answers to Problems

3.1

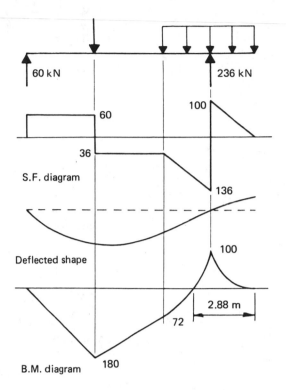

**Figure P3.6**

3.2

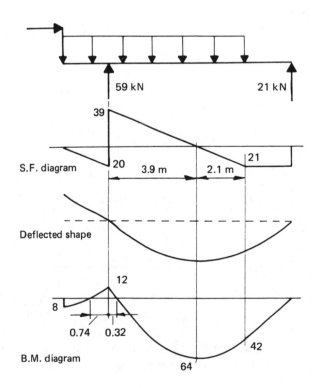

**Figure P3.7**

**3.3**

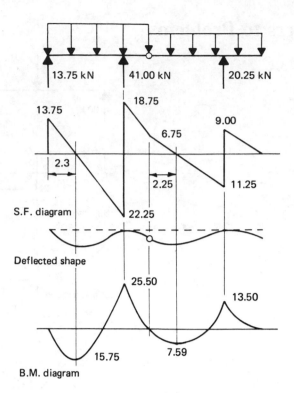

Figure P3.8

**3.4**

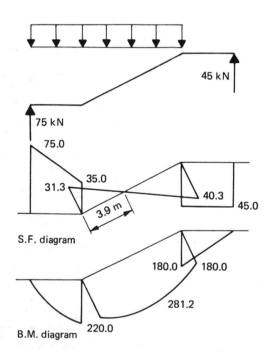

**Figure P3.9**  Thrusts: in AC = 0; in DB = 0; in CD varies from 20.1 kN tension at D to 15.6 kN compression at C

**3.5**

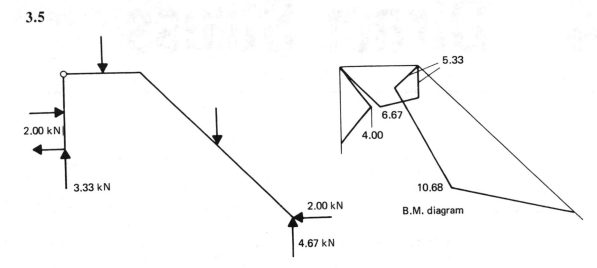

2.00 kN

3.33 kN

2.00 kN

4.67 kN

5.33

6.67

4.00

10.68

B.M. diagram

**Figure P3.10**

# 4 Direct Stress

## 4.1 Contents

Direct stress and strain ● Young's modulus of elasticity ● Change in length ● Lateral strain ● Poisson's ratio ● Change in volume ● Stresses in thin-walled cylinders.

## 4.2 The Fact Sheet

### (a) Direct (Normal) Stress

A force which acts normal to a surface causes stress which also acts normal to that surface. Provided that the force passes through the centroid of the surface, then the stress will be uniform over the whole surface. This type of stress is known as a *direct* stress ($\sigma$). The direct stress can be calculated from the expression

$$\sigma = \frac{\text{axial force } (P)}{\text{area of surface } (A)}$$

### (b) Direct Strain

A body subject to a direct stress will deform and be in a state of *strain*. The direct strain ($\epsilon$), which is measured in the same direction as the direct stress ($\sigma$), is given by the expression

$$\epsilon = \frac{\text{change in length of body } (\delta L)}{\text{original length of body } (L)}$$

where both $\delta L$ and $L$ are measured in the direction of the applied normal force ($P$).

## (c)  Stress–Strain Relationship

The direct stress ($\sigma$) and the direct strain ($\epsilon$) are related by Young's modulus of elasticity ($E$) in the expression

$$E = \frac{\text{direct stress } (\sigma)}{\text{direct strain } (\epsilon)}$$

## (d)  Lateral Strain

The straining of any elastic material will give rise to a change in lateral dimensions and, hence, to *lateral strains* in all directions at right angles to the direction of the applied force.

## (e)  Poisson's Ratio

For any elastic material, the direct (longitudinal) strain and the indirect lateral strain are related by Poisson's Ratio ($v$), which is given by the expression

$$v = -\frac{\text{lateral strain}}{\text{longitudinal strain}}$$

## (f)  Stresses in Thin-walled Cylinders

Thin-walled cylinders, such as pressure vessels, are subject to direct stresses along their length normal to all cross-sections of the cylinder. They are also subject to stresses which act circumferentially around the cylinder. These circumferential stresses are called *hoop stresses*. Although there are standard formulae for the calculation of both longitudinal and hoop stresses, they are best calculated by the application of fundamental principles, as illustrated by the examples in this chapter.

# 4.3  Symbols, Units and Sign Conventions

$A$ = area ($mm^2$)
$E$ = Young's modulus of elasticity ($kN/mm^2$)
$L$ = length (m)
$P$ = force (kN)
$\epsilon$ = strain (strain is dimensionless and therefore has no units)
$v$ = Poisson's ratio (dimensionless)
$\sigma$ = stress ($N/mm^2$)

Tensile stress and tensile strain are both taken as positive.

## 4.4 Worked Examples

### Example 4.1

A draw bar between a tractor and a trailer is made from a length of steel with a rectangular cross-section 100 mm by 12 mm. The load is transmitted to the bar via a pin through a 25 mm diameter hole at each end (see Figure 4.1).

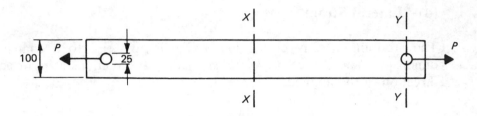

**Figure 4.1**

(a) If the load P in the bar is 100 kN, determine the stress at a section $(X–X)$ half-way along the bar, and the stress at a section $(Y–Y)$ through the bar at one of the pin positions.

(b) If the maximum permissible stress in the steel is 150 N/mm$^2$, determine the maximum load that can be taken by the bar.

### Solution 4.1

(a)

$$\text{Cross-sectional area of bar at } X–X = 100 \times 12 = 1200 \text{ mm}^2$$
$$\text{Stress at } X–X = \text{load/area} = 100\,000/1200 = \underline{83.33 \text{ N/mm}^2}$$

$$\text{Cross-sectional area of bar at } Y–Y = (100 - 25) \times 12 = 900 \text{ mm}^2$$
$$\therefore \text{ stress at } Y–Y = \text{load/area} = 100\,000/900 = \underline{111.11 \text{ N/mm}^2}$$

(b)
**The stress at $Y–Y$ governs the safe load that can be taken by the bar.**

$$\text{Permissible stress at } Y–Y = 150 \text{ N/mm}^2$$
$$\therefore \text{ Maximum load} = \text{stress} \times \text{area} = 150 \times 900 \times 10^{-3} = 135 \text{ kN}$$

**Alternatively, maximum load (by proportion) equals (150/111.11) × 100 = 135 kN.**

### Example 4.2

A laboratory vacuum flask consists of two long thin-walled glass tubes, each with wall thickness 1 mm, connected by thick diaphragms as shown in Figure 4.2. The space between the tubes is completely evacuated. Determine the maximum

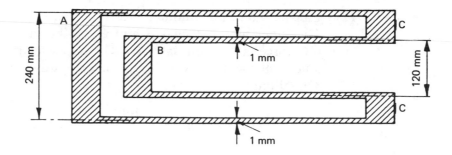

**Figure 4.2**

tensile and compressive stresses in the glass. (Normal atmospheric pressure = 0.101 N/mm².)

(Birmingham University)

To enable reference to be made to the diagram, the faces of the diaphragms have been lettered. Figure 4.3 shows the free body diagrams of parts of the flask obtained by taking a section through the inner and outer tubes. Note that, since the flask is evacuated, atmospheric pressure exerts inward-acting forces on the outer surfaces of the flask.

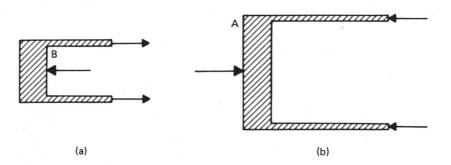

**Figure 4.3**  (a) Free body diagram of part of inner tube; (b) free body diagram of part of outer tube

*Solution 4.2*

(1)  Calculation of forces acting on faces B and A

$$\text{Area of face B} = \frac{\pi \times 119^2}{4} = 11\ 122.01\ \text{mm}^2$$

$\therefore$ Force on face B, acting to the left = area of face $\times$ pressure acting on it
$$= 11\ 122.01 \times 0.101 \times 10^{-3}$$
$$= 1.123\ \text{kN}$$

Total area of face A = $(\pi \times 241^2)/4 = 45\ 616.67\ \text{mm}^2$
$\therefore$ Force on face A, to the right, $= 45\ 616.67 \times 0.101 \times 10^{-3}$
$$= 4.607\ \text{kN}$$

(2) Calculation of longitudinal stresses

The inner tube between B and C will be in tension under the action of the force acting on face B. The cross-sectional area of glass in the inner tube $= \pi \times$ average diameter $\times$ wall thickness $= \pi \times 120 \times 1 = 376.99$ mm$^2$.

$$\therefore \text{Tensile stress} = \frac{\text{force}}{\text{area}} = \frac{1.123 \times 10^3}{376.99} = \underline{2.98 \text{ N/mm}^2}$$

The outer tube will be in compression under the action of the force on face A. The cross-sectional area of glass in the outer tube $= \pi \times 240 \times 1 = 753.98$ mm$^2$.

$$\therefore \text{Compressive stress} = \frac{\text{force}}{\text{area}} = \frac{4.607 \times 10^3}{753.98} = \underline{6.11 \text{ N/mm}^2}$$

(3) Calculation of hoop stresses

Figure 4.4 shows a half-section through the outer tube. The atmospheric pressure acts radially inwards, as indicated, giving rise to a resultant force $R_o$ which is resisted by forces $P_o$ acting on the faces D. Considering a 1 mm length of tube:

$$\text{force } R_o = \text{pressure} \times \text{projected area}$$
$$= (0.101) \times (241 \times 1) = 24.34 \text{ N}$$

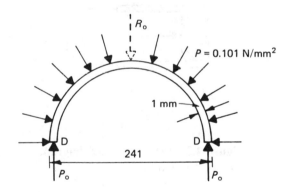

**Figure 4.4**  Outer tube

**Note that the resultant thrust on a curved surface equals the value of the pressure on the surface multiplied by the projection of the area on to a plane surface at right angles to the direction of action of the resultant thrust.**

$$\text{Area of one face D} = 1 \times 1 = 1 \text{ mm}^2$$
$$\therefore \text{Force } P_o = \text{area} \times \text{stress} = 1 \times \sigma_o$$

where $\sigma_o$ is the hoop stress in the outer tube. But $2P_o = R_o$.

$$\therefore 2\sigma_o = 24.34$$

$$\therefore \underline{\text{Compressive hoop stress in outer tube} = 12.17 \text{ N/mm}^2}$$

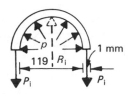

**Figure 4.5**  Inner tube

Figure 4.5 shows a half-section through the inner tube. In this case the atmospheric pressure gives rise to a tensile hoop stress in the tube.

Considering a 1 mm length of the inner tube:

$$\text{Force } R_i = 0.101 \times (119 \times 1) = 12.02 \text{ N}$$
$$\text{Force } P_i = \text{area} \times \text{stress} = 1 \times \sigma_i$$

where $\sigma_i$ is the hoop stress in the inner tube.

$$2P_i = R_i$$
$$\therefore 2\sigma_i = 12.02$$

$\therefore$ Tensile hoop stress in inner tube = 6.01 N/mm$^2$

**It should be noted that this solution for the hoop stresses assumes that the effect of the solid end plates may be ignored. Such an assumption is suggested by the question, which states that the tubes are 'long'. Note that in the given solution the actual external and internal diameters of the tubes have been used in the calculations. However, it is normally sufficiently accurate when analysing 'thin'-walled tubes to use the standard formulae:**

$$\textbf{longitudinal stress} = (p \times D)/4t$$

**and**

$$\textbf{hoop stress} = (p \times D)/2t$$

**where $p$ = the internal or external pressure;**
**$t$ = the thickness of the wall of the tube; and**
**$D$ = the *mean* diameter of the tube.**

### Example 4.3

The structure shown in Figure 4.6 consists of three uniform horizontal members AB, BC and CD. The members are pinned together at B and C, while the ends A and D are pinned to rigid foundations. Calculate the value of the horizontal force at B that will produce a displacement of 1.8 mm at B in the $x$ direction. What is then the displacement at joint C and the reactions at A and D? The members have the following cross-sectional areas:   AB = 1200 mm$^2$
BC = 1000 mm$^2$
CD = 1400 mm$^2$

(Assume that $E$ = 200 kN/mm$^2$.)

(Manchester University)

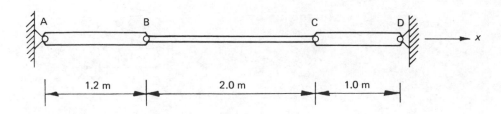

**Figure 4.6**

The overall distance between A and D is fixed at 4.2 m and can not change. Thus, the extension of AB by a force applied in the *x* direction at B must be equal to the combined shortening of BC and CD by the same force. Any force applied at B may be considered as comprising two parts: a force $P_1$ required to stretch AB and a force $P_2$ required to compress BCD. These forces are shown in the free body diagrams in Figure 4.7.

**Figure 4.7**

## Solution 4.3

(1)  Let: $P$ = the total force applied at B
$P_1$ = the component of $P$ stretching AB
$P_2$ = the component of $P$ compressing BCD
Then

$$\text{stress in AB} = \sigma = \text{force/area} = P_1/1200 \text{ kN/mm}^2$$
$$\text{strain in AB} = \epsilon = \sigma/E = P_1/(1200 \times 200)$$

and

$$\text{increase in length of AB} = \delta L = \epsilon \times L = (P_1 \times L)/(A \times E)$$
$$= (P_1 \times 1200)/(1200 \times 200) \text{ mm} = P_1/200 \text{ mm}$$

(2)  Similarly, decrease in length of BC

$$= (P_2 \times 2000)/(1000 \times 200) \text{ mm} = P_2/100 \text{ mm}$$

and

the decrease in length of CD

$$= (P_2 \times 1000)/(1400 \times 200) \text{ mm} = P_2/280 \text{ mm}$$

(3)  But the increase in length of AB equals the combined decrease in length of BC and CD.

$$\therefore P_1/200 = P_2/100 + P_2/280$$

Multiplying by 280:

$$1.4P_1 = 2.8P_2 + P_2$$
$$\therefore P_2 = 0.368P_1$$

The increase in length of AB is required to be 1.8 mm.

$$\therefore P_1/200 = 1.8.$$
$$\therefore P_1 = 360.00 \text{ kN}$$

and

$$P_2 = 0.368 \times P_1 = 0.368 \times 360 = 132.48 \text{ kN}$$
$$\therefore \text{Total force } P = P_1 + P_2 = 360.00 + 132.48 = \underline{492.48 \text{ kN}}$$

(4)  The displacement of joint C equals the shortening of CD

$$= P_2/280 = 132.48/280 = \underline{0.47 \text{ mm}}$$

The reaction at A equals the force in member AB = $\underline{360.00 \text{ kN}}$

The reaction at D equals the force in BCD = $\underline{132.48 \text{ kN}}$

## Example 4.4

A rigid gate ABC, pinned at its base A, is used to retain water in a long tank as shown in Figure 4.8. The gate is tied back to a foundation by steel bars BD which are spaced at 3 m centres along the length of the tank. The bars have a cross-sectional area of 300 mm$^2$. Determine the horizontal deflection of C when the gate retains a 5 m height of water.
(The density of water is 1000 kg/m$^3$. $E$ for steel is 200 kN/mm$^2$.)

(Birmingham University)

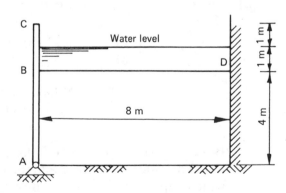

**Figure 4.8**

**The hydrostatic pressure exerted on the gate increases linearly from zero at the water surface to $wh$ at the base, where $w$ is the unit weight of water and $h$ is the depth of water retained (see Figure 4.9). The unit weight $w$ equals $\rho g$, where $\rho$ is the density of water and $g$ is the gravitational acceleration.**

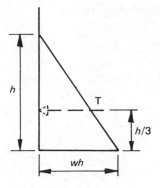

**Figure 4.9** Pressure distribution diagram

### Solution 4.4

(1)

$$\text{Maximum pressure on gate} = \rho gh = (1000 \times 9.81 \times 5)/1000$$
$$= 49.05 \text{ kN/m}^2$$
$$\therefore \text{Average pressure on gate} = 49.05/2 = 24.525 \text{ kN/m}^2$$

Each steel bar restrains a 3 m length of the gate.

Hydrostatic thrust on one 3 m length
$$= \text{average pressure} \times \text{wetted area of gate}$$
$$= 24.525 \times (3 \times 5)$$
$$= 367.88 \text{ kN (acting at } h/3 = \tfrac{5}{3} \text{ m above the base)}$$

**The resultant thrust ($T$) acts through the centroid of the pressure distribution diagram.**

(2)  Let $P$ equal the force in one bar. Then, taking moments about A,
($\Sigma M_A = 0$)
$$-367.88 \times \tfrac{5}{3} + P \times 4 = 0$$
$$\therefore P = 153.28 \text{ kN}$$

$$\text{extension of bar} = (PL)/(AE)$$
$$= (153.28 \times 8000)/(300 \times 200)$$
$$= 20.44 \text{ mm}$$

But end D of the bar is fixed; thus, end B of the bar moves 20.44 mm to the left. The gate AB is 'rigid' and thus may be assumed to act as a straight lever which can rotate about A. Thus, by proportion,

$$\text{movement of C} = \text{AC/AB} \times \text{movement of B}$$
$$= \tfrac{6}{4} \times 20.44 = \underline{30.66 \text{ mm}}$$

**Example 4.5**

A Demec gauge is used to measure the strain in a metal strip (width $B$ and thickness $t$) carrying an axial tensile force $F$.

(a)  If the dimensions $B$ and $t$, measured at the centre of the gauge length ($L$), are found to be 25.0 mm and 2.0 mm, respectively, and the measured strain for a force of 9.45 kN is $900 \times 10^{-6}$, determine the value of Young's modulus, $E_1$.

(b)  Closer examination of the strip shows it to be tapered, as shown in Figure 4.10, with the width varying from $b_1$ to $b_2$. By considering the extension of an element of length $\delta x$ at a distance $x$ from Q, show that the total extension over the gauge length ($L$) is given by;

$$\text{extension} = \frac{F \times L \times \log_e(b_1/b_2)}{t \times E \times (b_1 - b_2)}$$

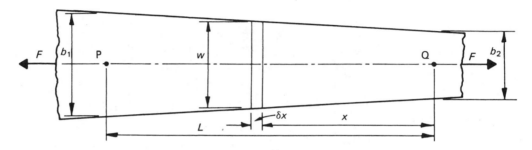

**Figure 4.10**

(c)  If the values of $b_1$ and $b_2$ are found to be 27.0 mm and 23.0 mm, respectively, determine a new value for the Young's modulus $E_2$. (The remaining data are given in part (a).)

(d)  Comment on the two answers obtained, $E_1$ and $E_2$.

(Nottingham University)

***Solution 4.5***

(a)

$$\text{Stress in strip, } \sigma = F/A = (9.45)/(25 \times 2) = 0.189 \text{ kN/mm}^2$$
$$\text{Strain, } \epsilon = 900 \times 10^{-6}$$
$$\therefore E_1 = \text{stress/strain} = \sigma/\epsilon = 0.189/(900 \times 10^{-6})$$
$$= \underline{210.00 \text{ kN/mm}^2}$$

(b)  The area ($A$) of cross-section at distance $x$ from Q is given by

$$A = w \times t = \{b_2 + (x/L)(b_1 - b_2)\} \times t$$
$$\therefore \text{Extension of element} = (F \times \delta x)/(A \times E)$$

$$= \frac{F \times \delta x}{\{b_2 + (x/L)(b_1 - b_2)\} \times t \times E}$$

107

and the total extension over the whole length of the strip will be

$$\delta L = \int_0^L \frac{F \, dx}{\{b_2 + (x/L)(b_1 - b_2)\} \times t \times E}$$

i.e.

$$\delta L = \frac{F}{t \times E} \int_0^L \frac{dx}{\{b_2 + (x/L)(b_1 - b_2)\}}$$

Let $U = \{b_2 + (x/L)(b_1 - b_2)\}$. Then

$$\frac{dU}{dx} = \frac{(b_1 - b_2)}{L}$$

and

$$\delta L = \frac{F}{t \times E} \int_0^L \frac{dU}{U} \times \frac{dx}{dU}$$

$$= \frac{F}{t \times E} \int_0^L \frac{dU}{U} \times \frac{L}{(b_1 - b_2)}$$

$$= \frac{F \times L}{t \times E \times (b_1 - b_2)} \int_0^L \frac{dU}{U}$$

$$= \frac{FL}{tE(b_1 - b_2)} \left[ \log_e U \right]_0^L$$

$$= \frac{FL}{tE(b_1 - b_2)} \left[ \log_e\{b_2 + (x/L)(b_1 - b_2)\} \right]_0^L$$

$$= \frac{FL}{tE(b_1 - b_2)} \left[ \log_e\{b_2 + L/L(b_1 - b_2)\} - \log_e b_2 \right]$$

$$= \frac{FL}{tE(b_1 - b_2)} \left[ \log_e b_1 - \log_e b_2 \right]$$

$$\therefore \delta L = \frac{FL \times \log_e(b_1/b_2)}{tE(b_1 - b_2)}$$

(c)  Substituting the values $b_1 = 27.0$ mm and $b_2 = 23.0$ mm:

$$\delta L = \frac{9.45 \times L \times \log_e(27.0/23.0)}{2.0 \times E_2 \times (27.0 - 23.0)}$$

$$\therefore \text{strain } \epsilon = \delta L/L = \frac{9.45 \times \log_e 1.1739}{2.0 \times E_2 \times 4.0}$$

$$\therefore E_2 = \frac{9.45 \times \log_e 1.1739}{2.0 \times 4.0 \times \epsilon}$$

But

$$\text{measured strain } \epsilon = 900 \times 10^{-6}$$

$$\therefore E_2 = \frac{9.45 \times \log_e 1.1739}{2.0 \times 4.0 \times 900 \times 10^{-6}}$$

$$= \underline{210.44 \text{ kN/mm}^2}$$

(d)   It is apparent that a slight variation in width does not have a significant effect on the value derived for Young's modulus. In this example a variation of $\pm 2.0$ mm (i.e. $\pm 8\%$) in width produces a variation of only 0.21% in the value of $E$.

## Example 4.6

A steel plate 200 mm $\times$ 400 mm $\times$ 5 mm thick is subjected to perpendicular tensile stresses of 140 N/mm$^2$ and 60 N/mm$^2$ in the plane of the plate, the 140 N/mm$^2$ stress acting on the shorter (200 mm $\times$ 5 mm) side.

Young's modulus of elasticity, $E = 200$ kN/mm$^2$.
Poisson's ratio, $\nu = 0.3$.

Determine the increase in the area and the decrease in the thickness of the plate.

(Polytechnic of Central London)

**No sketch accompanied this question. The first stage of any solution should be to sketch an appropriate diagram, as in Figure 4.11, and designate appropriate $X$, $Y$ and $Z$ axes.**

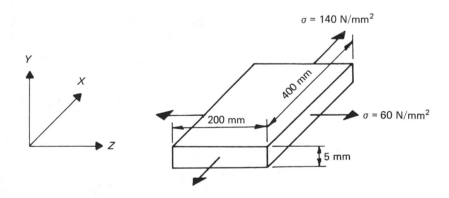

**Figure 4.11**

**It is convenient, and helps to avoid errors, if the solution is set out in a tabular form as follows. Note that the table is first set out using symbols and signs, the actual values being set out in a second table using the first table as a guide.**

(1)  Calculation of strains

| Strain | Due to | | |
|---|---|---|---|
| | $\sigma_x$ | $\sigma_y$ | $\sigma_z$ |
| $\epsilon_x$ | $+\sigma_x/E$ | $-\nu\sigma_y/E$ | $-\nu\sigma_z/E$ |
| $\epsilon_y$ | $-\nu\sigma_x/E$ | $+\sigma_y/E$ | $-\nu\sigma_z/E$ |
| $\epsilon_z$ | $-\nu\sigma_x/E$ | $-\nu\sigma_y/E$ | $+\sigma_z/E$ |

In this example:

| Strain | Due to | | | Total |
|---|---|---|---|---|
| | $\sigma_x$ | $\sigma_y$ | $\sigma_z$ | |
| $\epsilon_x$ | $+140/E$ | $0$ | $-0.3 \times 60/E$ | $+122/E$ |
| $\epsilon_y$ | $-0.3 \times 140/E$ | $0$ | $-0.3 \times 60/E$ | $-60/E$ |
| $\epsilon_z$ | $-0.3 \times 140/E$ | $0$ | $+60/E$ | $+18/E$ |

Thus, the strains are:

in the x direction, $\epsilon_x = +122/E = +122/(200 \times 10^3) = +610 \times 10^{-6}$
in the y direction, $\epsilon_y = -60/E = -300 \times 10^{-6}$
in the z direction, $\epsilon_z = +18/E = +90 \times 10^{-6}$

(2)  Calculation of increase in area

Increase in length = strain × original length
$= +610 \times 10^{-6} \times 400 = 0.244$ mm
Increase in width = $+90 \times 10^{-6} \times 200 = 0.018$ mm
∴ New area = $(400 + 0.244) \times (200 + 0.018) = 80\ 056.0$ mm$^2$
Original area = $400 \times 200 = 80\ 000.0$ mm$^2$

∴ Increase in area = 56.0 mm$^2$

(3)  Calculation of change in thickness

Change in thickness of plate = $\epsilon_y$ × plate thickness
$= -(300 \times 10^{-6}) \times 5$
$= -0.0015$ mm

i.e.

decrease in thickness = 0.0015 mm

## Example 4.7

(a)  Show that when a material is subjected to uniaxial tension or compression without lateral constraint, the change in volume is given by

$$V' = V \times (\sigma/E) \times (1 - 2\nu)$$

where  $V'$ = the change in volume due to the axial stress;
$V$ = the original volume before the stress was applied;
$\sigma$ = the axial stress applied;
$E$ = Young's modulus of elasticity;
$\nu$ = Poisson's ratio for the material.

(b) Hence show that, for a material which is stressed in the same manner as (a) and where Poisson's ratio is 0.5, there will be no change in volume.

(c) Determine the change in volume which will occur in a concrete cylinder of 250 mm diameter and 1000 mm length when it is subjected to an axial compression of 200 kN. Young's modulus $(E) = 210$ kN/mm$^2$ and Poisson's ratio $(\nu) = 0.2$.

(Coventry University)

### Solution 4.7

(a)
**First sketch a diagram as Figure 4.12.**

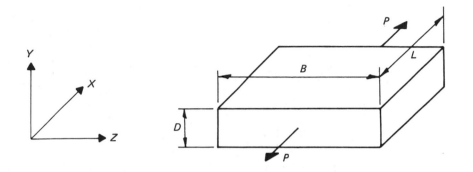

**Figure 4.12**

Longitudinal stress, $\sigma = P/BD$
Longitudinal strain, $\epsilon = +\delta L/L$

where $\delta L$ is the increase in length.

Lateral strain, $\epsilon_L = -\delta B/B$ and $-\delta D/D$

where $\delta B$ and $\delta D$ are the decreases in $B$ and $D$ respectively.

Original volume $= L \times B \times D$
Volume after loading and subsequent deformation
$$= (L + \delta L) \times (B - \delta B) \times (D - \delta D)$$
$$= L(1 + \epsilon)B(1 + \epsilon_L)D(1 + \epsilon_L)$$
$$= LBD(1 + \epsilon)(1 + 2\epsilon_L + \epsilon_L{}^2)$$

Expanding the brackets and neglecting second-order terms in $\epsilon$ and $\epsilon_L$,

volume after loading $= LBD(1 + \epsilon + 2\epsilon_L)$.

Thus,

$$\text{change in volume} = \text{final volume} - \text{original volume}$$
$$= LBD(1 + \epsilon + 2\epsilon_L) - LBD$$
$$= LBD(\epsilon + 2\epsilon_L)$$
$$= LBD\epsilon(1 - 2\nu)$$

since $\epsilon_L = -\nu\epsilon$.

Thus,
$$V' = V \times (\sigma/E) \times (1 - 2\nu)$$

(b)  If $v = 0.5$, then

$$\text{change in volume} = V \times (\sigma/E) \times (1 - 2 \times 0.5) = 0$$

(c)  Area of cross-section, $A = \pi d^2/4 = \pi \times 250^2/4$
$$= 49.087 \times 10^3 \text{ mm}^2$$
$$\therefore \text{ original volume} = 1000 \times 49.087 \times 10^3$$
$$= 49.087 \times 10^6 \text{ mm}^3$$

and

$$\text{compressive stress, } \sigma = P/A$$
$$= 200 \times 10^3/49.087 \times 10^3$$
$$= 4.074 \text{ N/mm}^2$$
$$\therefore \text{ Change in volume} = V \times (\sigma/E) \times (1 - 2v)$$
$$= 49.087 \times 10^6 \times (4.074/210 \times 10^3)$$
$$\times (1 - 2 \times 0.2)$$
$$= 571.37 \text{ mm}^3$$

## Example 4.8

A cylindrical reinforced concrete water tank sits on solid ground and is filled with water to a depth of 6 m. The internal diameter of the tank is 10 m. If the circumferential hoop stress is limited to 1.5 N/mm², determine the required thickness of the walls of the tank. (*Hint*  The pressure in the tank at depth $h$ is $\rho g h$, where $\rho = 1000$ kg/m³.)

(Coventry University)

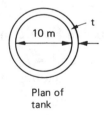

Plan of
tank

**Figure 4.13**

*Solution 4.8*

(1)  Maximum pressure ($p$) at base of wall of tank

$$= \rho g h \text{ (see Figure 4.14)}$$
$$= 1000 \times 9.81 \times 6$$
$$= 58860 \text{ N/m}^2$$
$$= 58.86 \times 10^{-3} \text{ N/mm}^2$$

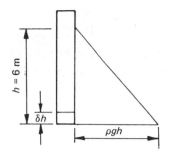

**Figure 4.14** Pressure distribution diagram

(2) **Consider a semicircular element of the wall at a level just above the base of the tank. Figure 4.15 shows the free body diagram of such an element of height δ$h$.**

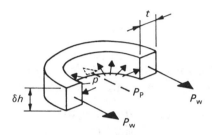

**Figure 4.15**

The water pressure ($p$) gives rise to a resultant force $P_P$, where

$P_P$ = water pressure × projected area of curved surface of tank
$\quad = (58.86 \times 10^{-3}) \times (10 \times 10^3 \times \delta h)$ N
$\quad = 588.60 \times \delta h$ N

The force $P_P$ is resisted by forces $P_w$ in the wall of the tank, as shown in Figure 4.15. If the hoop stress in the wall at base level is $\sigma_H$, then the force $P_w$ is given by

$P_w = \sigma_H \times$ area of cross-section of wall
$\quad = \sigma_H \times (\delta h \times t)$

But, for equilibrium, the two internal forces ($P_w$) must equal the force ($P_P$) due to the water pressure. Thus,

$$2 \times \sigma_H \times (\delta h \times t) = 588.60 \times \delta h$$

But $\sigma_H$ is limited to 1.5 N/mm²; thus the required value of $t$ is given by

$$2 \times 1.5 \times t = 588.60$$

$$\therefore t = 196.20 \text{ mm}$$

113

# 4.5 Problems

**4.1** A vertical steel hanger ABC is 2.0 m long and carries a load of 135 kN at the lower end, as shown in Figure P4.1. The upper length AB is 50 mm diameter and

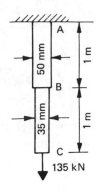

**Figure P4.1**

the lower length BC is 35 mm diameter. If Young's modulus of elasticity ($E$) is 200 kN/mm$^2$ and Poisson's ratio ($\nu$) is 0.3, calculate:

  (i)   the total extension;
 (ii)  the vertical movement of B;
(iii)  the change in lateral dimensions of AB and BC;
(iv)  the change in volume of the hanger.

(Coventry University)

**4.2** A *rigid* bar PQR of length 6 m is pinned to a foundation at P and supported by a steel cable QS of cross-sectional area 270 mm$^2$ at Q (see Figure P4.2). QS is of length 2.5 m and its end S is pinned to a foundation. The bar PQR is subjected to a uniformly distributed load of 21 kN/m along its length. Assuming that the displacements of the bar PQR are small, calculate the deflection of R. ($E$ = 200 kN/mm$^2$.)

(Birmingham University)

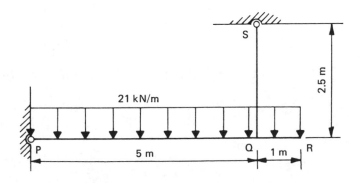

**Figure P4.2**

**4.3** A cube is made from a material whose modulus of elasticity is $10^5$ N/mm$^2$ and Poisson's ratio is 0.25. Two opposite faces of the cube are subjected to a compressive stress of 80 N/mm$^2$. Determine the percentage change in volume of the cube.

(University of Westminster)

**4.4** A long cylindrical pressure vessel of internal diameter 400 mm and length 10.0 m has plane ends as shown in Figure P4.3. The vessel contains a fluid at a pressure of 3.1 N/mm$^2$ and the wall thickness of the vessel is 4 mm.

(a) Determine:

  (i) the total fluid thrust on one end;
  (ii) the longitudinal stress in the wall of the vessel; and
  (iii) the hoop stress in the wall of the vessel.

(b) If the ends of the vessel are secured to the body of the vessel by 12 bolts as indicated and if the tensile stress in the bolts is not to exceed 165 N/mm$^2$, determine the size (diameter) of bolts required.

(Coventry University)

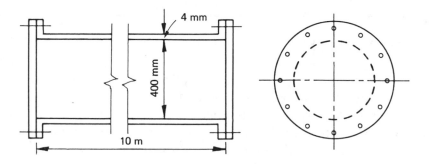

**Figure P4.3**

# 4.6 Answers to Problems

**4.1** (i) 1.04 mm; (ii) 0.34 mm; (iii) −0.005 mm and −0.007 mm; (iv) 540 mm$^3$

**4.2** 4.20 mm

**4.3** 0.04%

**4.4** (a) (i) 389.56 kN; (ii) 77.50 N/mm$^2$; (iii) 155.00 N/mm$^2$. (b) 15.83 mm

# 5 Bending Stress

## 5.1 Contents

Neutral axis • Calculation of second moments of area • Elastic section modulus • Theory of bending • Moments of resistance • Bending stresses in symmetrical and assymetrical sections • Compound sections • Use of flange plates to strengthen beams.

## 5.2 The Fact Sheet

### (a) Bending Stress

The effect of a bending moment applied to a cross-section of a beam is to induce a state of stress across that section. These stresses are known as bending stresses and they act normally to the plane of the cross-section.

Bending stresses vary linearly across a section, with maximum value (compression or tension) at the outer fibres of the beam and with zero value at the level of the *neutral axis*.

### (b) Neutral Axis

The *neutral axis* is the axis of the cross-section of a beam at which both the bending strain and bending stress are zero. The neutral axis passes through the centroid of the cross-section.

### (c) Equations of Bending

The fundamental equations that govern the bending of a beam or other structural member are given by

$$\frac{\sigma}{y} = \frac{M}{I} = \frac{E}{R}$$

where    $\sigma$ = the bending stress in a layer of fibres distance $y$ from the neutral axis;

        $M$ = the bending moment at the section under consideration;

$I$ = the second moment of area of the cross-section taken about the neutral axis;

$E$ = Young's modulus of elasticity; and

$R$ = the radius of curvature at the section under consideration.

### (d)   Second Moment of Area

The *second moment of area* of a cross-section about the neutral axis is a geometrical property of that cross-section and can be obtained from tables of standard values or can be calculated.

For a rectangular section of breadth $b$ and depth $d$, the second moment of area ($I_{CC}$) of that section about an axis through the centroid is given by:

$$I_{CC} = \frac{bd^3}{12}$$

### (e)   Parallel Axis Theorem

The second moment of area ($I_{XX}$) of a cross-section about an $X$–$X$ axis which is parallel to and at a distance $h$ from an axis through its own centroid is given by

$$I_{XX} = I_{CC} + Ah^2$$

where   $I_{CC}$ = the second moment of area about the centroidal axis; and

$A$ = the area of the cross-section.

### (f)   Elastic Section Modulus

The elastic section modulus ($Z$) of a section is the second moment of area ($I$) about the axis of bending, divided by the distance ($y_{max}$) from the neutral axis to the furthermost fibre of the section. That is,

$$Z = \frac{I}{y_{max}}$$

If the neutral axis is not at the mid-height of a section, that section will have two section moduli—one relating to the compression face and one relating to the tensile face.

### (g)   Moment of Resistance

The moment of resistance ($M$) of a beam at a particular section is given by

$$M = \sigma_{max} \times Z$$

where   $\sigma_{max}$ = the maximum permissible bending stress; and

$Z$ = the elastic section modulus of that section.

## 5.3  Symbols, Units and Sign Conventions

$A$ = area (mm$^2$)

$h$ = the distance between an axis through the centroid of a section and any parallel axis (mm)

$I_{CC}$ = the second moment of area of a section about an axis through the centroid of its own cross-sectional area (mm$^4$ or cm$^4$)

$I_{XX}$ = the second moment of area of a section about an $X$–$X$ axis (mm$^4$ or cm$^4$)

$M$ = moment of resistance or bending moment (kNm)

$\sigma$ = bending stress (N/mm$^2$)

$y$ = the distance from the neutral axis to the level at which the bending stress is being calculated (mm)

$y_{max}$ = the distance from the neutral axis to the top or bottom face of the beam (mm)

$Z$ = elastic section modulus of a section about an $X$–$X$ axis (mm$^3$ or cm$^3$)

$z$ = a distance used for the determination of the position of the neutral axis of a section ($y$ is commonly used for this purpose, but $z$ is used in this text, to avoid confusion with the term $y$ as defined above) (mm)

## 5.4  Worked Examples

### Example 5.1

A column section is built up by connecting together two 305 mm × 102 mm channels, using battens (see Figure 5.1). If the column is to have equal second moments of area about its two axes of symmetry, calculate the spacing ($2p$) required between the two channels. (Note that the battens do not contribute to the second moments of area of the section.)

(Birmingham University)

Properties of one channel:  $A = 60$ cm$^2$
$I_{xx} = 8210$ cm$^4$
$I_{yy} = 500$ cm$^4$

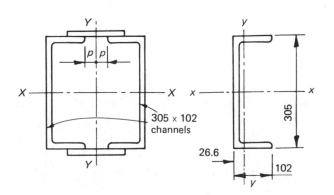

**Figure 5.1**

118

*Solution 5.1*

(1) **The second moment of area of the compound section about the X–X axis will be equivalent to the sum of the second moments of area of the two channels about their own x–x axes. This is because the x–x axes of the two channels and the X–X axis of the compound section are colinear.**

$$\therefore I_{XX} = 2 \times I_{xx}$$
$$= 2 \times 8210 = \underline{16\ 420\ \text{cm}^4}$$

(2) **The y–y axes of the individual channels are not colinear with the Y–Y axis of the compound section; thus, the parallel axes theorem must be used to determine the value of $I_{YY}$ for each channel about the Y–Y axis.**

Let the distance between the Y–Y axis of the compound section and the y–y axis of a single channel be $h$; then, for one channel,

$$I_{YY} = I_{yy} + Ah^2$$
$$= 500 + 60 \times h^2$$

Therefore, for the two channels,

$$I_{YY} = 2(500 + 60h^2)$$
$$= 1000 + 120 \times h^2$$

But, as the column is to have equal second moments of area about both the X–X and the Y–Y axes, then

$$I_{YY} = I_{XX}$$

that is,

$$1000 + 120h^2 = 16\ 420$$
$$\therefore h = [(16\ 420 - 1000)/120]^{1/2}$$
$$= 11.336\ \text{cm}$$
$$= 113.36\ \text{mm}$$

But, from the geometry of the section, $h$ is also given by

$$h = (p + 102 - 26.60)$$
$$\therefore (p + 102 - 26.60) = 113.36$$
$$\therefore p = 113.36 - 102 + 26.60$$
$$= 37.96$$

Hence,

$$\underline{2p = 75.92\ \text{mm}}$$

Note that the area and second moment of area of the channel section are given in cm units; thus, the derived value of 11.336 for $h$ is in cm units, the other dimensions and $p$ being in mm. Values of $I$ (and $Z$) are normally quoted in cm units in standard tables, whereas calculations (as in future examples) are more usually executed in mm units. The reader must take care to be consistent in the use of units.

## Example 5.2

The beam section shown in Figure 5.2 is formed of a channel section and a universal beam section welded together as shown. The cross-sectional areas ($A$) and the relevant second moments of areas ($I_{CC}$) about the centroidal axes of each section are as tabulated. The distance 26.5 mm is the distance from the top of the compound beam to the centroid of the channel section.

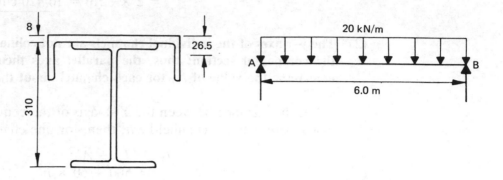

| Part | $A$ (cm$^2$) | $I_{CC}$ (cm$^4$) |
|---|---|---|
| Channel | 38.0 | 264 |
| Universal beam | 60.8 | 9500 |

**Figure 5.2**

Calculate the maximum tensile and compressive bending stresses caused by the 20 kN/m load on a span of 6.0 m, as indicated.

(Liverpool University)

## Solution 5.2

(1) To locate the neutral axis of the compound section and to determine the value of $I_{XX}$, moments of area are taken about the bottom face of the section. ($z$ is measured from the bottom face of the compound section.)

| Part | Area ($A$) (mm$^2$) | $z$ (mm) | $Az$ (mm$^3 \times 10^3$) | $h(z - \bar{z})$ (mm) | $Ah^2$ (mm$^4 \times 10^6$) | $I_{CC}$ (mm$^4 \times 10^6$) |
|---|---|---|---|---|---|---|
| Channel | 3800 | 291.50 | 1107.70 | 84.00 | 26.81 | 2.64 |
| Universal beam | 6080 | 155.00 | 942.40 | 52.50 | 16.76 | 95.00 |
| | 9880 | | 2050.10 | | 43.57 | 97.64 |

$$\bar{z} = \frac{\Sigma Az}{\Sigma A} = \frac{2050.1 \times 10^3}{9880} = \underline{207.5 \text{ mm}}$$

$$\therefore I_{XX} = 43.57 + 97.64 = \underline{141.21 \times 10^6 \text{ mm}^4}$$

120

(2)  The maximum bending stresses will occur at mid-span, where the bending moment is a maximum.

Maximum bending moment (at mid-span)
$$= wL^2/8$$
$$= (20 \times 6^2)/8$$
$$= 90.00 \text{ kNm} = 90.00 \times 10^6 \text{ Nmm}$$

Bending stress is given by

$$\sigma = \frac{M}{I} \times y$$

and will have maximum tensile value at the bottom of the section, where $y$ is 207.5 mm.

$$\therefore \text{ Maximum tensile bending stress} = \frac{90.00 \times 10^6}{141.21 \times 10^6} \times 207.5$$

$$= \underline{132.25 \text{ N/mm}^2}$$

and maximum compressive stress at the top of the section

$$= \frac{90.00 \times 10^6}{141.21 \times 10^6} \times (318.00 - 207.50)$$

$$= \underline{70.43 \text{ N/mm}^2}$$

## Example 5.3

The fibre-reinforced concrete I-beam shown in Figure 5.3(a) is simply supported over a span of 8 m and carries a superimposed load as indicated in Figure 5.3(b). The density of the concrete is 2400 kg/m³.

Determine the maximum value of bending moment in the beam due to:

(i)   self-weight only;
(ii)  imposed load only.

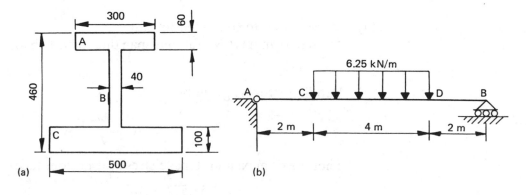

**Figure 5.3**

Calculate the maximum tensile bending stress due to a combination of self-weight and imposed loading.

(University of Portsmouth)

**For ease of reference, the flanges and web of the section have been lettered.**

### Solution 5.3

(1) To locate the neutral axis and calculate $I_{XX}$, moments of area are taken about the bottom face of the section ($z$ is measured from the bottom face of the section).

| Part | Area ($A$) ($\text{mm}^2 \times 10^3$) | $z$ (mm) | $Az$ ($\text{mm}^3 \times 10^3$) | $h$ (mm) | $Ah^2$ ($\text{mm}^4 \times 10^6$) | $I_{CC}(bd^3/12)$ ($\text{mm}^4 \times 10^6$) |
|------|------|------|------|------|------|------|
| A | 18 | 430 | 7740 | 264.50 | 1259.28 | 5.40 |
| B | 12 | 250 | 3000 | 84.50 | 85.68 | 90.00 |
| C | 50 | 50 | 2500 | 115.50 | 667.01 | 41.67 |
| | 80 | | 13 240 | | 2011.97 | 137.07 |

$$\bar{z} = \frac{\Sigma Az}{\Sigma A} = \frac{13\,240 \times 10^3}{80 \times 10^3} = 165.50$$

$$\therefore I_{XX} = 2011.97 + 137.07 = \underline{2149.04 \times 10^6 \text{ mm}^4}$$

(2) Bending moment due to self-weight
Volume of beam = area × length = $(80\,000 \times 10^{-6}) \times 1 = 0.08 \text{ m}^3/\text{m}$
$\therefore$ weight = $0.08 \times (2400 \times 9.81 \times 10^{-3}) = 1.884 \text{ kN/m}$
$\therefore$ Maximum bending moment at mid-span = $wL^2/8$
$= 1.884 \times 8^2/8 = \underline{15.072 \text{ kNm}}$

(3) Bending moment due to the imposed load
Reaction $V_A = V_B = (6.25 \times 4)/2 = 12.5 \text{ kN}$
Maximum B.M. (at mid-span) = $(12.5 \times 4) - (6.25 \times 2 \times 2/2)$
$= \underline{37.500 \text{ kNm}}$

(4) Stresses due to combined loading
Total maximum B.M. (at mid-span due to both loadings) = $15.072 + 37.50$
$= \underline{52.572 \text{ kNm}}$

Bending stress is given by

$$\sigma = \frac{M}{I} \times y$$

Therefore, maximum tensile stress (at bottom face)

$$\sigma = \frac{52.572 \times 10^6}{2149.04 \times 10^6} \times 165.5 = \underline{4.05 \text{ N/mm}^2}$$

**If asked for, the maximum compressive stress would be calculated as follows:**

maximum compressive stress (at top face)

$$\sigma = \frac{52.572 \times 10^6}{2149.04 \times 10^6} \times (460 - 165.5) = \underline{7.20 \text{ N/mm}^2}$$

### Example 5.4

Figure 5.4 shows a 254 mm × 102 mm universal beam which is strengthened by a 102 mm × 12 mm plate fixed to its lower flange. The compound beam is used as a cantilever 4 m long, with the plate extending 1 m from the support. Details of the unreinforced universal beam are:

$$254 \text{ mm} \times 102 \text{ mm U.B.} \qquad \text{Area} = 36.2 \text{ cm}^2 \qquad I_{XX} = 4008 \text{ cm}^4$$

If the maximum allowable stress in tension and compression is 160 N/mm$^2$, calculate the allowable value of $W$ for the loading shown.

(University of Hertfordshire)

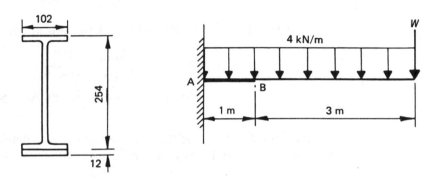

**Figure 5.4**

**The bending moment is a maximum at the support A, where the beam is reinforced by the addition of the flange plate. Hence, it is necessary to first analyse the compound section of the beam to determine the moment of resistance of the beam at A.**

### Solution 5.4

(1) To locate the neutral axis and calculate $I_{XX}$ ($z$ measured from the bottom face of the section)

| Part | Area ($A$) (mm$^2$) | $z$ (mm) | $Az$ (mm$^3$ × 10$^3$) | $h$ (mm) | $Ah^2$ (mm$^4$ × 10$^6$) | $I_{CC}$ (mm$^4$ × 10$^6$) |
|------|------|------|------|------|------|------|
| Universal beam | 3620 | 139 | 503.18 | 33.61 | 4.09 | 40.08 |
| Plate | 1224 | 6 | 7.34 | 99.39 | 12.09 | 0.01 |
| | 4844 | | 510.52 | | 16.18 | 40.09 |

$$\bar{z} = \frac{\Sigma A z}{\Sigma A} = \frac{510.52 \times 10^3}{4844} = 105.39 \text{ mm}$$

$$\therefore I_{XX} = 16.18 + 40.09 = \underline{56.27 \times 10^6 \text{ mm}^4}$$

(2)  To determine the moment of resistance at A

Moment of resistance  $M = \sigma_{max} \times Z = \sigma_{max} \times \dfrac{I}{y_{max}}$

Considering the compression face of the beam:

**In this case the compression face is at the bottom of the beam since the beam is acting as a cantilever.**

$$M = 160 \times \frac{56.27 \times 10^6}{105.39} \times 10^{-6} = \underline{85.43 \text{ kNm}}$$

Considering the tensile face of the beam:

**In this case the tensile face is at the top of the beam.**

$$M = 160 \times \frac{56.27 \times 10^6}{(266.00 - 105.39)} \times 10^{-6} = \underline{56.06 \text{ kNm}}$$

$$\therefore \text{ Moment of resistance at A} = \underline{56.06 \text{ kNm}}$$

**Note that the lower of the two possible values determines the working value for the moment of resistance of the beam.**

(3)  To determine the moment of resistance at B

**B is a critical section, since at this section the externally applied moment must be carried by the universal beam alone. The stress situation at this section must be checked.**
The universal beam is symmetrical; hence, the stresses at the top and bottom flanges will be equal. The moment of resistance is, therefore, given by:

$$M = \sigma_{max} \times Z = \sigma_{max} \times \frac{I}{y_{max}}$$

$$= 160 \times \frac{40.08 \times 10^6}{127} \times 10^{-6} = \underline{50.49 \text{ kNm}}$$

(4)  To determine the maximum allowable load
Taking moments about B of forces to the right of B, the bending moment at B is given by:

$$M_B = (4 \times 3) \times \tfrac{3}{2} + W \times 3$$
$$= (18 + 3W) \text{ kNm}$$

But, as already calculated, the moment of resistance of the beam at B is 50.49 kNm. Hence, equating the bending moment to the moment of resistance,

$$18 + 3W = 50.49$$
$$\therefore W = 10.83 \text{ kN}$$

Similarly, the bending moment at A is given by

$$M_A = (4 \times 4) \times \tfrac{4}{2} + W \times 4$$
$$= (32 + 4W) \text{ kNm}$$

and the moment of resistance of the compound beam at A is 56.06 kNm. Hence, equating the bending moment to the moment of resistance,

$$32 + 4W = 56.06$$
$$\therefore W = 6.02 \text{ kN}$$

$$\therefore \text{Maximum allowable value for } W = \underline{6.02 \text{ kN}}$$

**Note that the lower of the two possible values for $W$ is the critical solution.**


## Example 5.5

The simply supported beam AB in Figure 5.5 is required to carry a vertical load of 15 kN at any point on the span. The beam has a rectangular hollow box cross-section of outside dimensions 100 mm × 75 mm which has two 65 mm × 6 mm plates welded to it over the middle 2 m section of the span. There is one plate welded to the top surface and one plate welded to the bottom surface as shown.

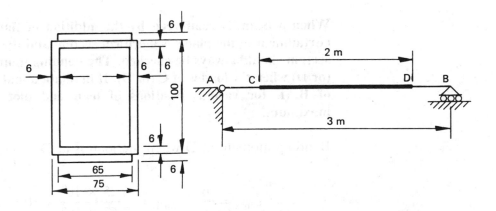

**Figure 5.5**

Determine the maximum bending stress and the maximum curvature in the beam as the 15 kN load moves across the span. ($E = 200 \text{ kN/mm}^2$.)

(Manchester University)

**The neutral axis of the section is at mid-depth, since the section is symmetrical.**

125

**Solution 5.5**

(1) To determine $I_{XX}$ for the strengthened and unstrengthened section

$$I_{XX} \text{ for the box section alone } = I_{\text{gross section}} - I_{\text{internal void}}$$
$$= (BD^3)/12 - (bd^3)/12$$
$$= (75 \times 100^3 - 63 \times 88^3)/12$$
$$= 2.67 \times 10^6 \text{ mm}^6$$

For one plate:

$$I_{XX} = I_{CC} + Ah^2 = 65 \times 6^3/12 + 65 \times 6 \times 53^2 = 1.10 \times 10^6 \text{ mm}^4$$
$$\therefore I_{XX} \text{ for strengthened section} = (2.67 + 2 \times 1.10) \times 10^6$$
$$= \underline{4.87 \times 10^6 \text{ mm}^4}$$

(2) At mid-span

The bending moment is a maximum at mid-span when the load is at mid-span and is of value

$$M = WL/4 = 15 \times \tfrac{3}{4} = 11.25 \text{ kNm}$$

Therefore, when the load is at mid-span, the maximum bending stress is given by

$$\sigma_{\text{max}} = \frac{M}{I} \times y_{\text{max}} = \frac{11.25 \times 10^6}{4.87 \times 10^6} \times 56 = 129.36 \text{ N/mm}^2$$

and curvature $1/R$ is given by

$$\frac{1}{R} = \frac{M}{EI} = \frac{11.25 \times 10^6}{(200 \times 10^3)(4.87 \times 10^6)} = 0.0116 \times 10^{-3} \text{ mm}^{-1}$$

(3) At the end of the flange plates (i.e. at C or D, 0.5 m from the supports)

**When a beam is reinforced by the addition of flange plates, the point of curtailment of the plates is a critical section and the state of stress at such a section should always be checked. The bending moment is a maximum at C (or D) when the load is at C (or D). If in doubt about this, calculate the value of B.M. for various positions of load and plot a graph to locate the maximum.**

Bending moment at D when the load is at D $= V_B \times 0.5$
$$= 12.5 \times 0.5 = 6.25 \text{ kNm}$$

$$\therefore \sigma_{\text{max}} = \frac{M}{I} \times y_{\text{max}} = \frac{6.25 \times 10^6}{2.67 \times 10^6} \times 50 = 117.04 \text{ N/mm}^2$$

and curvature is given by

$$\frac{1}{R} = \frac{M}{EI} = \frac{6.25 \times 10^6}{(200 \times 10^3)(2.67 \times 10^6)} = 0.0117 \times 10^{-3} \text{ mm}^{-1}$$

$\therefore$ Maximum bending stress = $\underline{129.36 \text{ N/mm}^2}$ (tension and compression)

maximum curvature = $\underline{0.0117 \times 10^{-3} \text{ mm}^{-1}}$

**Example 5.6**

A horizontal simply supported uniform beam spans 2 m and is made from the section shown in Figure 5.6. A vertical point load of 25 kN can be placed anywhere on the span. If the maximum stress in tension or compression is not allowed to exceed 160 N/mm², determine whether there is a region along the beam where it would be unsafe to place the load. Determine the limits of this region if it exists.

(Manchester University)

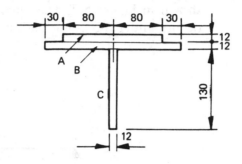

**Figure 5.6**

*Solution 5.6*

(1) To locate the neutral axis and calculate $I_{XX}$ ($z$ is measured from the bottom face of the section)

| Part | Area ($A$) (mm²) | $z$ (mm) | $Az$ (mm³ × 10³) | $h$ (mm) | $Ah^2$ (mm⁴ × 10⁶) | $I_{CC}(bd^3/12)$ (mm⁴ × 10⁶) |
|------|------|------|------|------|------|------|
| A | 1920 | 148 | 284.16 | 26.33 | 1.33 | 0.02 |
| B | 2640 | 136 | 359.04 | 14.33 | 0.54 | 0.03 |
| C | 1560 | 65 | 101.40 | 56.67 | 5.01 | 2.20 |
| | 6120 | | 744.60 | | 6.88 | 2.25 |

$$\bar{z} = \frac{\Sigma Az}{\Sigma A} = \frac{744.60 \times 10^3}{6120} = 121.67 \qquad I_{XX} = 6.88 + 2.25 = 9.13 \times 10^6$$

(2) To determine the moment of resistance

$$Z = \frac{I}{y_{max}} = \frac{9.13 \times 10^6}{121.67}$$

$$= 75.04 \times 10^3 \text{ mm}^3$$

Hence,

$$M = \sigma_{max} \times Z = 160 \times 75.04 \times 10^3 \times 10^{-6}$$
$$= 12.01 \text{ kNm}$$

**Note that the elastic section modulus based on the distance to the bottom face of the beam has been used. This is the smaller of the two elastic modulii and, hence,**

127

gives the critical value for the moment of resistance in an example such as this, in which the maximum permissible values of the tensile and compressive stresses are equal in magnitude.

(3) To locate the limits of the region where the load should not be placed

**The maximum B.M. in a beam due to a single concentrated load will always be at the point of application of the load.**

Consider the load at a distance $x$ from the right-hand end of the beam, as shown in Figure 5.7. The maximum bending moment in the beam, which occurs under the point load, is given by

$$M = V_B \times x = \frac{25(2 - x)}{2} \times x = 25x - 12.5x^2$$

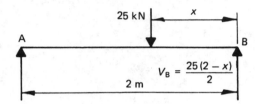

**Figure 5.7**

Equating the bending moment to the moment of resistance,

$$25x - 12.5x^2 = 12.01$$

i.e.

$$12.5x^2 - 25x + 12.01 = 0$$

from which,

$$x = 0.8 \text{ m or } 1.2 \text{ m}$$

Thus, it is not safe to place the load further than 0.8 m from either end, and the central region of the beam (a length of 1.2–0.8 m, i.e. 0.4 m) constitutes the unsafe region.


## Example 5.7

A prismatic beam has the cross-sectional shape shown in Figure 5.8. It is subjected to a bending moment about the $X$–$X$ axis (i.e. the axis passing through the centre of area Q of the section) in such a way that compressive stresses are induced in the top flange.

Determine a suitable value for the width of the top flange, $b$, if the maximum allowable stresses in tension and compression are 200 N/mm² and 150 N/mm², respectively. What would be the bending moment that the beam could then safely carry?

(Manchester University)

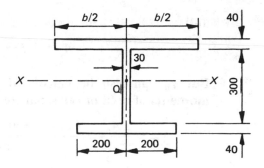

**Figure 5.8**

This question involves the design of the most economic section (i.e. one in which the maximum allowable stresses in both tension and compression are attained simultaneously).

*Solution 5.7*

(1)  To determine $b$

If the maximum tensile stress equals 200 N/mm² and the maximum compressive stress simultaneously equals 150 N/mm², then the moment of resistance of the section is given by

$$M = I \times \frac{200}{y_B}$$

and

$$M = I \times \frac{150}{y_T}$$

where  $y_B$ = the distance from the neutral axis to the bottom face of the beam (i.e. the face in tension) and
$y_T$ = the distance from the neutral axis to the top face of the beam (i.e. the face in compression).

Hence,

$$I \times \frac{200}{y_B} = I \times \frac{150}{y_T}$$

$$\therefore \frac{200}{y_B} = \frac{150}{y_T}$$

$$\therefore y_B = y_T \times \frac{200}{150} = 1.333 y_T$$

But

$$y_B + y_T = 380$$
$$\therefore 2.333 y_T = 380$$
$$y_T = 162.9 \text{ mm}$$

and

$$y_B = 1.333 \times 162.9 = 217.1 \text{ mm}$$

But $y_B$ can also be calculated (as in the previous examples) by taking moments of area of the separate parts of the section about the bottom face.

$$y_B = \frac{\Sigma Az}{\Sigma A} = \frac{(40b \times 360) + (300 \times 30 \times 190) + (40 \times 400 \times 20)}{40b + (300 \times 30) + (40 \times 400)}$$

$$\therefore 217.1 = \frac{(14.4b + 1710 + 320) \times 10^3}{(0.04b = 9 + 16) \times 10^3}$$

$$8.68b + 5427.50 = 14.4b + 2030.00$$
$$5.72b = 3397.5$$

$$\therefore \underline{b = 594 \text{ mm}}$$

(2) To determine $I_{XX}$

| Part | Area ($A$) ($\text{mm}^2 \times 10^3$) | $h$ (mm) | $Ah^2$ ($\text{mm}^4 \times 10^6$) | $I_{CC}$ ($\text{mm}^4 \times 10^6$) |
|------|------|------|------|------|
| Top flange | 23.76 | $360.0 - 217.1 = 142.9$ | 485.19 | 3.17 |
| Web | 9.00 | $217.1 - 190.0 = 27.1$ | 6.61 | 67.50 |
| Bottom flange | 16.00 | $217.1 - 20.0 = 197.1$ | 621.57 | 2.13 |
| | | | 1113.37 | 72.80 |

$$I_{XX} = 1113.37 + 72.80 = 1186.17 \times 10^6 \text{ mm}^4$$

(3) To determine the moment of resistance

$$M = I \times \frac{\sigma_{max}}{y_{max}} = 1186.17 \times 10^6 \times \frac{200}{217.1}$$

$$= 1093 \times 10^6 \text{ Nmm} = \underline{1093 \text{ kNm}}$$

## Example 5.8

A beam carries a central point load of $W$ kN on a simply supported span of 6 m. The section consists of a 178 mm $\times$ 102 mm joist of area 2740 $\text{mm}^2$ and second moment of area $15.2 \times 10^6 \text{ mm}^4$. A single 100 mm $\times$ 10 mm plate can be welded onto the bottom flange as shown in Figure 5.9. Calculate the maximum load $W$ and the minimum length of plate required if the allowable stress is 165 $\text{N/mm}^2$.

(University of Hertfordshire)

**Since the value of the bending moment in the beam reduces from a maximum at the mid-span to zero at the supports, the reinforcing flange plate is not required over the entire length. Figure 5.10 has been added to show the bending moment diagram with the values derived in the following calculations.**

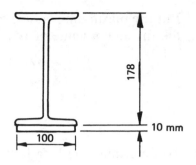

**Figure 5.9**

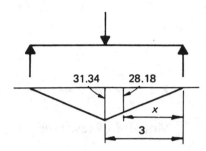

**Figure 5.10** B.M. diagram

## Solution 5.8

(1) To locate the neutral axis and calculate $I_{XX}$ for the beam with the additional flange plate

| Part | Area $(A)$ $(\text{mm}^2)$ | $z$ $(\text{mm})$ | $Az$ $(\text{mm}^3 \times 10^3)$ | $h$ $(\text{mm})$ | $Ah^2$ $(\text{mm}^4 \times 10^6)$ | $I_{CC}$ $(\text{mm}^4 \times 10^6)$ |
|------|------|------|------|------|------|------|
| Joist | 2740 | 99.00 | 271.26 | 25.13 | 1.73 | 15.20 |
| Plate | 1000 | 5.00 | 5.00 | 68.87 | 4.74 | 0.01 |
| | 3740 | | 276.26 | | 6.47 | 15.21 |

$$\bar{z} = \frac{\Sigma Az}{\Sigma A} = \frac{276.26 \times 10^3}{3740} = 73.87 \text{ mm}$$

$$I_{XX} = 6.47 + 15.21 = 21.68 \times 10^6 \text{ mm}^4$$

(2) To determine maximum allowable value of $W$

The moment of resistance is given by

$$M = \sigma_{\max} \times Z = \sigma_{\max} \times \frac{I}{y_{\max}} = 165 \times \frac{21.68 \times 10^6}{(188 - 73.87)} \times 10^{-6} = 31.34 \text{ kNm}$$

**Note that the value of $y_{\max}$ from the neutral axis to the *top* face has been used to calculate the moment of resistance, since this is the larger of the two $y_{\max}$ values and gives the lesser (maximum allowable) value for $M$.**

131

But the maximum bending moment due to the load $W = WL/4$. Therefore, the maximum value of $W$ is given by

$$WL/4 = W \times 6/4 = 31.34 \text{ kNm}$$

i.e.

$$W = \underline{20.89 \text{ kN}}$$

(3) To determine the length of plate required

For the joist acting alone,

$$I_{XX} = 15.20 \times 10^6 \text{ mm}^4$$

and

$$y_{max} = 178/2 = 89 \text{ mm}$$

$$\therefore Z = \frac{I}{y_{max}} = \frac{15.20 \times 10^6}{89} = 170.79 \times 10^3 \text{ mm}^3$$

$$\therefore \text{Moment of resistance } M = \sigma_{max} \times Z = 165 \times (170.79 \times 10^3) \times 10^{-6}$$
$$= 28.18 \text{ kNm}$$

Therefore, the plate may be curtailed at a section, distance $x$ from the end of the beam, where the value of the bending moment is equal to the moment of resistance of the joist (i.e. 28.18 kNm). By reference to Figure 5.10,

$$\frac{x}{3} = \frac{28.18}{31.34}$$

$$\therefore x = 2.70 \text{ m}$$

and

$$\text{length of plate required} = 6 - 2 \times 2.70 = \underline{0.60 \text{ m}}$$

## Example 5.9

The box-section beam of Figure 5.11(a) is simply supported at B and C and has the cross-section shown in Figure 5.11(b).

(a) Draw the shearing force and bending moment diagrams, giving peak values.
(b) Determine the maximum compressive and tensile bending stresses occurring in the beam.
(c) If the maximum value of bending stress (tension or compression) can be up to 120 N/mm², determine the maximum length to which the centre span BC can be increased.

(University of Nottingham)

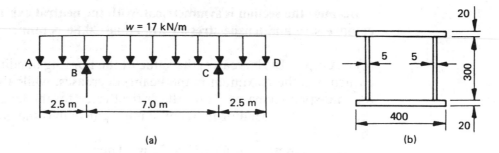

**Figure 5.11**

## Solution 5.9

(a)
The shearing force and bending moment diagrams are drawn in Figure 5.12. **(Since the beam is symmetrical, the construction should be straightforward, with a minimum of calculations.)**

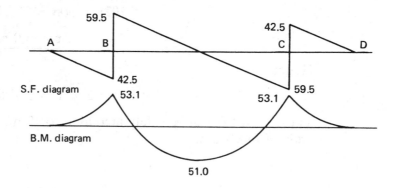

**Figure 5.12**

(b)
To calculate $I_{XX}$

**The section is symmetrical; thus, the neutral axis is at mid-depth.**

| Part | Area ($A$) (mm$^2$) | $h$ (mm) | $Ah^2$ (mm$^4 \times 10^6$) | $I_{CC}$ ($bd^3/12$) (mm$^4 \times 10^6$) |
|---|---|---|---|---|
| Top flange | 8000 | 160.00 | 204.80 | 0.27 |
| Bottom flange | 8000 | 160.00 | 204.80 | 0.27 |
| 1st web | | | 0.00 | 11.25 |
| 2nd web | | | 0.00 | 11.25 |
| | | | 409.60 | 23.04 |

$$I_{XX} = 409.60 + 23.04 = 432.64 \times 10^6 \text{ mm}^4$$

$$\therefore \sigma_{max} = \frac{M}{I} \times y_{max} = \frac{53.1 \times 10^6}{432.64 \times 10^6} \times 170 = \underline{20.86 \text{ N/mm}^2}$$

133

**Because the section is symmetrical, with the neutral axis at mid-depth, both the compressive and tensile stresses will equal 20.86 N/mm².**

(c)   As span BC increases, the value of the hogging bending moment at B and C (which was the maximum in the beam) decreases, while the bending moment at the mid-span increases. Thus, the critical value for the length of span BC is when the value of the bending moment at mid-span equals the moment of resistance of the beam.

Let the length of span BC = $2x$ m. Then

$$\text{bending moment at mid-span} = (102 \times x) - (17 \times 6 \times \tfrac{6}{2})$$
$$= 102x - 306 \text{ kNm}$$

$$\text{Moment of resistance of beam} = I_{XX} \times \frac{\sigma_{max}}{y_{max}}$$

$$= 432.64 \times 10^6 \times \frac{120}{170} \times 10^{-6} = 305.39 \text{ kNm}$$

Equating the mid-span moment to the moment of resistance,

$$102x - 306 = 305.39$$

$$\therefore x = 6.0 \text{ m}$$

Hence,

$$\text{span BC} = 2 \times x = 2 \times 6 = 12 \text{ m}$$

Thus, span BC can be safely increased to 12 m, and the supports moved to the ends of the beam.

## 5.5  Problems

**5.1**   A beam has the cross section shown in Figure P5.1 and resists a maximum hogging moment of 132 kNm. Determine whether the section is over- or understressed if the permissible stresses are 110 N/mm² in tension and 90 N/mm² in compression.

(Liverpool University)

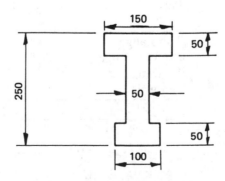

**Figure P5.1**

**5.2**  The T-beam shown in Figure P5.2 is to be used as a simply supported beam spanning 2 m. Determine the maximum uniformly distributed load that the beam can carry if the compressive and tensile bending stresses are not to exceed 150 N/mm² and 200 N/mm², respectively.

Draw the bending stress distribution for the section at the centre of the span when subjected to this maximum load.

(Salford University)

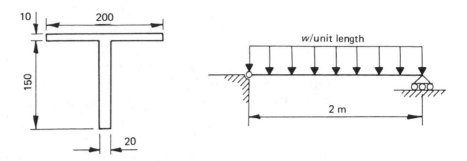

**Figure P5.2**

**5.3**  Figure P5.3 shows the cross-section of a structural member. This section is to be used as a beam spanning 10 m between simple supports and carrying a superimposed load of 3 kN/m.

Determine the maximum stresses at the top and bottom of the section due to the self-weight and the superimposed load.

Take the density of the material in the section as 2200 kg/m³.

(University of Hertfordshire)

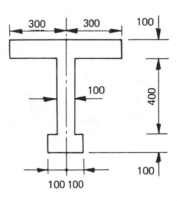

**Figure  P5.3**

**5.4**  A universal beam 406 mm deep, area 68.4 cm² and second moment of area 18 626 cm⁴ is fixed to a channel with web thickness 8.1 mm, area 36 cm², second moment of area 162.6 cm⁴ and centroid 1.86 cm from the back of the web to form a compound beam as shown in Figure P5.4.

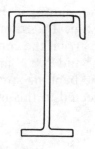

**Figure P5.4**

The beam carries a uniformly distributed load of 30 kN/m on a simply supported span of 5 m. If the allowable bending stress is 160 N/mm², calculate the maximum additional central point load that can be carried.

(University of Hertfordshire)

5.5 The beam ABCD is simply supported at B and C. It carries vertical uniformly distributed loads of 130 kN/m (B to C) and 50 kN/m (A to B and C to D), as shown in Figure P5.5(a). The member ABCD is a steel universal beam (610 × 229 × 101), as shown in Figure P5.5(b). Over the length EF (see Figure P5.5a) the U.B. is strengthened by the welding of 10 mm thick plates to the flanges, as shown in Figure P5.5(c).

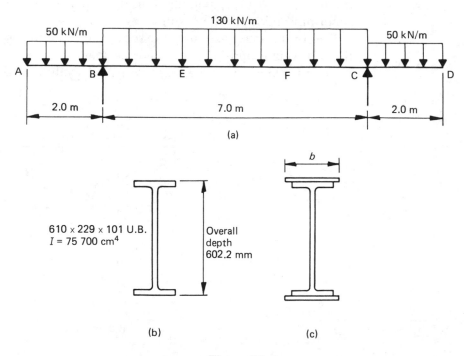

**Figure P5.5**

(a) Determine the values of the reaction forces at B and C and draw the shearing force and bending moment diagrams, indicating peak values.

(b) Show that if the maximum bending stress is limited to ± 165N/mm², the U.B. alone is sufficient to carry the bending moment at B and C.

(c)   Determine the width of flange plates (*b*) required over the central part of BC and the minimum length (EF) of the plates.

(Nottingham University)

(*Hint*   Although the density of the U.B. is quoted as 101 kg/m, the self-weight of the beam should be neglected in the calculations. Also note that the overall depth of the universal beam is 602.2 mm—the 610 mm is the nominal size only.)

**5.6**   A compound gantry girder is fabricated from a channel section welded to a universal beam section as indicated in the diagram of Figure P5.6. Determine the second moment of area *I* of the section about the neutral axis.

If the maximum permissible stress in tension is 200 N/mm² and in compression is 120 N/mm², determine the maximum allowable bending moments for sagging and hogging actions.

(Sheffield University)

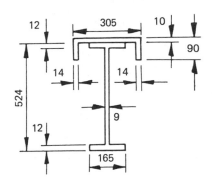

**Figure P5.6**

# 5.6   Answers to Problems

**5.1**   Overstressed in compression at 131.02 N/mm². Understressed in tension at 107.20 N/mm².

**5.2**   Maximum load = 49.84 kN/m, i.e. 99.68 kN total load. The stress diagram is linear from a maximum compressive stress of 99 N/mm² at the top face to a maximum tensile stress of 200 N/mm² at the bottom face.

**5.3**   Maximum compressive stress at top face = 3.18 N/mm²
Maximum tensile stress at bottom face = 5.62 N/mm²

**5.4**   Maximum additional point load = 55.81 kN.

**5.5**   (a)   The shearing force and bending moment diagrams are shown in Figure P5.7.
(b)   Moment of resistance of the U.B. alone = 414.83 kNm. Thus, the bending moment at B and C can be carried.
(c)   Width (*b*) = 297 mm. Length (EF) = 4.16 m.

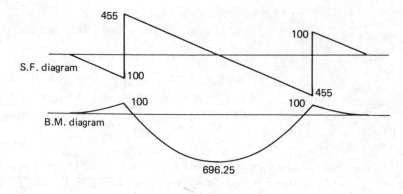

**Figure P5.7**

**5.6** B.M. based on maximum tensile stress (bottom face) = 311.84 kNm
B.M. based on maximum compressive stress (top face) = 378.63 kNm
($I_{XX}$ of compound section = 557.25 × 10⁶ mm⁴)

# 6 Combined Bending and Direct Stress

## 6.1 Contents

The determination of the stresses in a section due to direct axial loads and externally applied moments acting together ● The effect of eccentric loadings on structural sections and on foundations ● Stress distribution below foundations ● The effect of prestressing on concrete sections.

## 6.2 The Fact Sheet

### (a) Single-axis Bending

The maximum stresses in a section subject to direct axial loading together with an externally applied moment acting about either the $X$–$X$ axis or the $Y$–$Y$ axis are given by

$$\sigma = \frac{P}{A} \pm \frac{M_X}{Z_X}$$

or

$$\sigma = \frac{P}{A} \pm \frac{M_Y}{Z_Y}$$

respectively, where

$A$ = the area of the cross-section under load;
$P$ = the direct axial load;
$M_X$ = the external applied moment about the $X$–$X$ axis;
$M_Y$ = the external applied moment about the $Y$–$Y$ axis;
$Z_X$ = the elastic modulus of the section about the $X$–$X$ axis; and
$Z_Y$ = the elastic modulus of the section about the $Y$–$Y$ axis.

### (b)  Biaxial Bending

The maximum stresses in a section subject to direct axial loading together with
externally applied moments acting about two mutually perpendicular axes ($X$–$X$
and $Y$–$Y$) are given by

$$\sigma = \frac{P}{A} \pm \frac{M_X}{Z_X} \pm \frac{M_Y}{Z_Y}$$

### (c)  Eccentrically Loaded Sections

If a load $P$ acts at an eccentricity $e$ from the centroidal axis of a section, then that
loading is equivalent to:

  (i)   A direct load $P$ acting axially through the centroid of the section, together
        with
  (ii)  a moment of value $P \times e$ acting about the axis of bending which passes
        through the centroid of the section.

## 6.3  Symbols, Units and Sign Conventions

$A$ = the area of the cross-section under load (mm$^2$)
$e_x$ = the eccentricity of a load measured perpendicular to the $Y$ axis (mm)
$e_y$ = the eccentricity of a load measured perpendicular to the $X$ axis (mm)
$M_X$ = the external applied moment about the $X$–$X$ axis (kNm)
$M_Y$ = the external applied moment about the $Y$–$Y$ axis (kNm)
$P$ = the direct axial load (kN)
$Z_X$ = the elastic modulus of the section about the $X$–$X$ axis (mm$^3$)
$Z_Y$ = the elastic modulus of the section about the $Y$–$Y$ axis (mm$^3$)

In this chapter compressive stresses will be taken as positive. In the type of
problem to be solved the stresses are predominantly compressive and the sign
convention is chosen with this fact in mind.

## 6.4  Worked Examples

### Example 6.1

Figure 6.1 shows the cross-section of a beam made from aluminium alloy. The
origin of coordinates is located at the centroid. Compute the values of $I_{XX}$ and
$I_{YY}$.

  At a particular cross-section the beam is subjected simultaneously to (a) a
tension of 50 kN acting through the centroid, (b) a bending moment about the
$X$–$X$ axis of value 0.8 kNm which causes tension along the face AB and (c) a

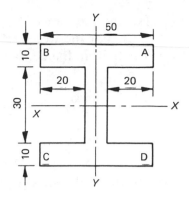

**Figure 6.1**

bending moment about the *Y–Y* axis of value 0.5 kNm which causes tension along the face BC.

Compute the stress on the cross-section at each of the four corners A, B, C and D due to the given loading, on the assumption that the behaviour is linear-elastic.

(Cambridge University)

**The section is symmetrical, the two axes of symmetry passing through the centroid and being colinear with the specified *X–X* and *Y–Y* axes. With reference to the last sentence of the question, the reader should appreciate that all the examples in this book are based on the assumption that materials behave in a linear-elastic manner.**

*Solution 6.1*

(1)   To calculate $Z_X$

| Part | Area ($A$) (mm$^2$) | $h$ (mm) | $Ah^2$ (mm$^4 \times 10^6$) | $I_{CC}$ ($bd^3/12$) (mm$^4 \times 10^6$) |
|------|------|------|------|------|
| Top flange | 500 | 20 | 0.200 | 0.004 17 |
| Web | 300 | 0 | 0.000 | 0.022 50 |
| Bottom flange | 500 | 20 | 0.200 | 0.004 17 |
| | 1300 | | 0.400 | 0.030 84 |

$$\therefore I_{XX} = 0.400 + 0.030\ 84 = \underline{0.431 \times 10^6 \text{ mm}^4}$$

$$\therefore Z_X = \frac{I_{XX}}{y_{max}} = \frac{0.431 \times 10^6}{25} = \underline{17.24 \times 10^3 \text{ mm}^3}$$

(2)   To calculate $Z_Y$

**The *Y* axis passes through the centroid of each of the three parts of the section; thus, there are no $Ah^2$ terms involved in the calculation.**

141

For the top flange, $bd^3/12 = 10 \times 50^3/12 = 0.104\ 17 \times 10^6$
For the web, $bd^3/12 = 30 \times 10^3/12 = 0.002\ 50 \times 10^6$
For the bottom flange, $bd^3/12 = 10 \times 50^3/12 = \underline{0.104\ 17 \times 10^6}$
$$0.210\ 84 \times 10^6$$

$$\therefore I_{YY} = \underline{0.211 \times 10^6\ \text{mm}^4}$$

$$\therefore Z_Y = \frac{I_{YY}}{x_{\max}} = \frac{0.211 \times 10^6}{25} = \underline{8.44 \times 10^3\ \text{mm}^3}$$

(3)  To determine the stresses on the cross-section

  (i)  Stress due to the direct axial tensile load of 50 kN:

  $$\frac{P}{A} = -\frac{50 \times 10^3}{1300} = -38.46\ \text{N/mm}^2 \text{ (i.e. a uniform tensile stress)}$$

  (ii)  Maximum stress due to $M_X$:

  $$\frac{M_X}{Z_X} = \pm \frac{0.8 \times 10^6}{17.24 \times 10^3} = \pm 46.40\ \text{N/mm}^2 \text{ (tensile on face AB and}$$
  $$\text{compressive on face CD)}$$

  (iii)  Maximum stress due to $M_{YY}$:

  $$\frac{M_Y}{Z_Y} = \pm \frac{0.5 \times 10^6}{8.44 \times 10^3} = \pm 59.24\ \text{N/mm}^2 \text{ (tensile on face BC and}$$
  $$\text{compressive on face AD)}$$

  Hence, the total stresses at the corners are given by:

at A, $\sigma_A = \dfrac{P}{A} - \dfrac{M_X}{Z_X} + \dfrac{M_Y}{Z_Y}$

$$= -38.46 - 46.40 + 59.24 = -25.62\ \text{N/mm}^2 \text{ (tensile)}$$

Similarly:

  at B, $\sigma_B = -38.46 - 46.40 - 59.24 = -144.10\ \text{N/mm}^2$ (tensile)
  at C, $\sigma_C = -38.46 + 46.40 - 59.24 = -51.30\ \text{N/mm}^2$ (tensile)
  at D, $\sigma_D = -38.46 + 46.40 + 59.24 = +67.18\ \text{N/mm}^2$ (compressive)

**The nature of the stresses produced by the axial load and the moments on the various faces are determined by a careful study of the question and the diagram. The reader should be able to visualise the effect each moment has on the cross-section (i.e. whether a moment causes tension or compression on any face or corner).**

## Example 6.2

A short box member carries an eccentric load $P$ as shown in Figure 6.2. The box member is 80 mm × 80 mm with a uniform wall thickness of 10 mm. If the value of the load is 72 kN, determine the values of the maximum tensile and

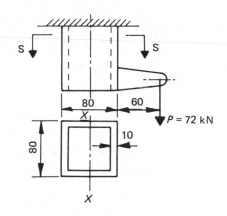

**Figure 6.2**

compressive stresses on a cross-sectional plane (such as S–S). Show the stress distribution on a diagram.

(Nottingham University)

**The eccentric load has an effect equivalent to that produced by a moment of value $P \times e$ (i.e. $72 \times (60 + 40)$ kNmm = 7.2 kNm) about the centroidal axis of the cross-section together with a direct axial tensile load of 72 kN. This equivalent loading is shown in Figure 6.3. $X$–$X$ is the centroidal axis, which for reference purposes has been added to the original Figure 6.2.**

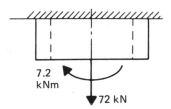

**Figure 6.3**

*Solution 6.2*

(1)   The stress on the section is given by

$$\sigma = \frac{P}{A} \pm \frac{Pe}{Z_X}$$

where area $A = (80 \times 80) - (60 \times 60) = 2800$ mm$^2$

eccentricity $e = 60 + 40 = 100$ mm

$y_{max} = 80/2 = 40$ mm

$I_{XX} = (80 \times 80^3)/12 - (60 \times 60^3)/12 = 2.33 \times 10^6$ mm$^4$

$Z_X = I_{XX}/y_{max} = 2.33 \times 10^6/40 = 58.25 \times 10^3$ mm$^3$

143

$$\therefore \sigma = -\frac{72 \times 10^3}{2.8 \times 10^3} \pm \frac{7.2 \times 10^6}{58.25 \times 10^3} = \underline{-149.32 \text{ N/mm}^2} \text{ (tensile)}$$

$$\text{or} \ \underline{+97.90 \text{ N/mm}^2} \text{ (compressive)}$$

(2)  The stress distribution across the section will be linear, varying from a maximum tensile stress of 149.32 N/mm² at the face nearest to the load $P$ to a maximum compressive stress of 97.90 N/mm² at the face furthest from the load. This is shown in Figure 6.4.

**Figure 6.4**

## Example 6.3

Figure 6.5 shows a short column which carries a single 10 kN compressive load eccentric to both axes. The column is of square hollow cross-section with external dimensions 150 mm × 150 mm and with a wall thickness of 10 mm. Determine the value of the stress at the extreme fibres at the positions marked a, b, c and d.

(University of Portsmouth)

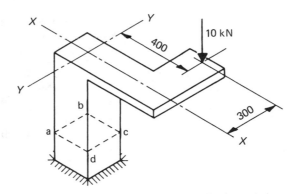

**Figure 6.5**

**The load is eccentric to both axes; thus, the equivalent loading will be:**

(i)   **a single direct compressive load of 10 kN acting at the centroid of the cross-section;**

(ii)  **a moment of value 10 × 300 kNmm = 3 kNm about the X–X axis of the cross-section and producing maximum compression on face bc;**

(iii) **a moment of value 10 × 400 kNmm = 4 kNm about the Y–Y axis of the cross-section and producing maximum compression on face cd.**

## Solution 6.3

The stresses at the corners are given by

$$\sigma = \frac{P}{A} \pm \frac{P \times e_y}{Z_X} \pm \frac{P \times e_x}{Z_Y}$$

where $P = 10$ kN

$A = (150 \times 150) - (130 \times 130) = 5600$ mm$^2$

$Pe_y = 10 \times 300 = 3000$ kNmm $= 3$ kNm

$y_{max} = 150/2 = 75$ mm

$Pe_x = 10 \times 400 = 4000$ kNmm $= 4$ kNm

$I_{XX} = (150 \times 150^3/12) - (130 \times 130^3/12) = 18.387 \times 10^6$ mm$^4$

$I_{YY} = I_{XX} = 18.387 \times 10^6$ mm$^4$

$Z_X = Z_Y = I_{XX}/y_{max} = 18.387 \times 10^6/75 = 245.16 \times 10^3$ mm$^3$

$$\therefore \text{stress } \sigma = \frac{10 \times 10^3}{5600} \pm \frac{3 \times 10^6}{245.16 \times 10^3} \pm \frac{4 \times 10^6}{245.16 \times 10^3}$$

$$= 1.786 \pm 12.237 \pm 16.316$$

$\therefore$ stress at a $= 1.786 - 12.237 - 16.316 = -26.77$ N/mm$^2$ (tensile)

at b $= 1.786 + 12.237 - 16.316 = -2.29$ N/mm$^2$ (tensile)

at c $= 1.786 + 12.237 + 16.316 = +30.34$ N/mm$^2$ (compressive)

at d $= 1.786 - 12.237 + 16.316 = +5.87$ N/mm$^2$ (compressive)

## Example 6.4

A cantilever of length 1 m has the cross-section shown in Figure 6.6 and carries vertical and horizontal forces of 1 kN and 200 kN, respectively, at its free end. Locate the position of the neutral axis of the section at the support under this loading and determine the maximum tensile stress on the section.

(University of Westminster)

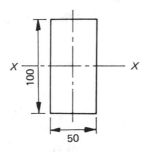

**Figure 6.6**

Figure 6.7 has been added to show the original loading on the cantilever and Figure 6.8 to show the equivalent loading at the fixed end.

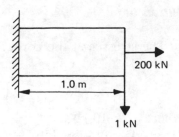

**Figure 6.7**

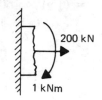

**Figure 6.8**  Equivalent loading

## Solution 6.4

(1)  The maximum bending moment in the cantilever will be at the fixed end and of value

$$M = 1 \times 1 = 1 \text{ kNm}$$

The elastic section modulus for a solid rectangular section is given by

$$Z_X = bd^2/6 = 50 \times 100^2/6 = 0.0833 \times 10^6 \text{ mm}^3$$

$$\therefore \text{ Bending stress } \sigma = \pm \frac{M_X}{Z_X} = \pm \frac{1 \times 10^6}{0.0833 \times 10^6} = \pm 12.00 \text{ N/mm}^2$$

$$\text{Direct stress} = P/A$$
$$= -200 \times 10^3/(50 \times 100) = -40.00 \text{ N/mm}^2$$

The total stress will be given by

$$\sigma = \frac{P}{A} \pm \frac{M_X}{Z_X}$$

i.e.

$\sigma = -40.00 - 12.00 = -52.00 \text{ N/mm}^2$ (tensile at the top face)

and

$\sigma = -40.00 + 12.00 = -28.00 \text{ N/mm}^2$ (tensile at the bottom face)

Thus,

$$\text{maximum tensile stress} = \underline{52.00 \text{ N/mm}^2}$$

(2) To locate the neutral axis

By reference to Figure 6.9, the neutral axis is seen to be external to the section and at a distance $x$ below the bottom face, where $x$ is given by

$$\frac{x}{(100 + x)} = \frac{28.00}{52.00}$$

$$\therefore x(52.00 - 28.00) = 2800$$
$$\therefore x = 2800/24 = 116.67 \text{ mm}$$

i.e. the neutral axis is 116.67 mm below the bottom face of the cantilever.

**Note that since this location is external to the section, the term 'neutral axis' does not have its usual physical meaning.**

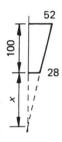

**Figure 6.9** Stress distribution diagram

## Example 6.5

Figure 6.10 shows the plan view of a concrete foundation which is 1 m thick and weighs 25 kN/m³. The two column loads are 300 kN and 800 kN in the positions shown. Assuming the base pressure to be uniform across the width and to vary linearly from end to end, determine the maximum and minimum pressures on the base.

(University of Westminster)

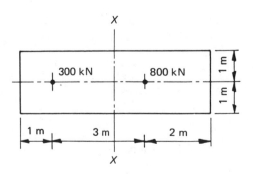

**Figure 6.10**

Figure 6.11 has been added to show the loads acting on the foundation and the distributed reactive earth pressure acting at the foundation ground interface.

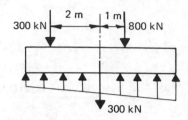

**Figure 6.11**

## Solution 6.5

(1)  The self-weight of the foundation acts through the centroid of the plan section and is of value $P = (6 \times 2 \times 1) \times 25 = 300$ kN.

(2)  The 300 kN load is equivalent to a direct load at the centroid of value $P = 300$ kN, together with a moment about the $X$–$X$ axis which will produce compressive stresses at the left-hand end of the foundation and is of value $M_X = 300 \times 2 = 600$ kNm.

(3)  The 800 kN load is equivalent to a direct load at the centroid of value $P = 800$ kN, together with a moment about the $X$–$X$ axis which will produce tensile stresses at the left-hand end of the foundation and is of value $M_X = 800 \times 1 = 800$ kNm.

(4)  The total equivalent loading is thus

   (i)  A direct compressive load at the centroid:

$$P = 300 + 300 + 800 = 1400 \text{ kN}$$

   (ii)  A moment about the $X$ axis through the centroid and of value $M_X = 800 - 600 = 200$ kNm, producing tensile stresses at the left-hand end of the foundation.

**This equivalent loading is shown in Figure 6.12.**

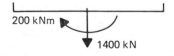

**Figure 6.12**

The pressure (stress) on the base of the foundation is given by

$$\sigma = \frac{P}{A} \pm \frac{M_X}{Z_X}$$

where

$$A = 6 \times 2 = 12 \text{ m}^2$$

and

$$Z_X = bd^2/6 = 2 \times 6^2/6 = 12 \text{ m}^3$$

$$\therefore \sigma = \frac{1400}{12} \pm \frac{200}{12}$$

$$= 116.67 \pm 16.67 \text{ kN/m}^2$$

$\therefore$ Minimum compressive pressure (stress) = $\underline{100.00 \text{ kN/m}^2}$

and

maximum compressive pressure = $\underline{133.34 \text{ kN/m}^2}$

## Example 6.6

A prismatic beam column AB is subjected to a compressive point load $W$ which acts parallel to its longitudinal axis as shown in Figure 6.13(a). The cross-section of the beam column is shown in Figure 6.13(b). If the load $W$ can be placed anywhere along the line of symmetry PQ on this section, determine its limiting positions such that there is no tensile stress present in the beam column.

(Manchester University)

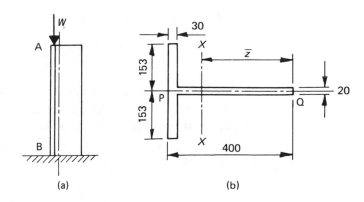

**Figure 6.13**

The section will bend about an $X$–$X$ axis through its centroid. This axis has been added to Figure 6.13(b). If the load acts at an eccentricity of $e$, measured from the centroid, then the equivalent loading is:
 (i)   a direct compressive force at the centroid of value $W$;
(ii)   a moment about the $X$ axis of value $W \times e$.

**Solution 6.6**

(1) To locate the centroid and calculate $I_{XX}$ and $Z_X$

| Part | Area ($A$) (mm$^2$) | $z$ (from Q) (mm) | $Az$ (mm$^3 \times 10^3$) | $h$ (mm) | $Ah^2$ (mm$^4 \times 10^6$) | $I_{CC}(bd^3/12)$ (mm$^4 \times 10^6$) |
|------|------|------|------|------|------|------|
| Flange | 9180 | 385 | 3534 | 89.28 | 73.17 | 0.69 |
| Web | 7400 | 185 | 1369 | 110.72 | 90.72 | 84.42 |
| | 16 580 | | 4903 | | 163.89 | 85.11 |

$$\bar{z} = \frac{\Sigma Az}{\Sigma A} = \frac{4903 \times 10^3}{16\,580} = 295.72$$

$$\therefore I_{XX} = (85.11 + 163.89) \times 10^6 = 249.00 \times 10^6 \text{ mm}^4$$

For the outside edge of the flange (i.e. at P):

$$Z_X = \frac{I_{XX}}{y_{max}} = \frac{249.00 \times 10^6}{(400 - 295.72)} = 2.388 \times 10^6 \text{ mm}^3$$

For the outside edge of the web (i.e. at Q):

$$Z_X = \frac{I_{XX}}{y_{max}} = \frac{249.00 \times 10^6}{295.72} = 0.842 \times 10^6 \text{ mm}^3$$

(2) To determine stresses

The maximum stresses are given by:

$$\sigma = \frac{W}{A} \pm \frac{M_X}{Z_X}$$

Therefore, the stress at P is given by

$$\sigma = \frac{W}{16\,580} \pm \frac{W \times e}{2.388 \times 10^6}$$

Tension will occur at P when the load acts at an eccentricity ($e$) in the direction of Q and if the resulting tensile bending stress at P is greater than the direct stress. For there to be no tensile stress at P, the limiting position of the load is given by equating the above expression to zero, with the second term negative—i.e.

$$W\left\{ \frac{1}{16\,580} - \frac{e}{2.388 \times 10^6} \right\} = 0$$

i.e.

$$\underline{e = 144.03 \text{ mm measured from the centroid towards Q}}$$

Similarly, for there to be no tensile stress at Q, the limiting value of $e$ is given by

$$W\left\{ \frac{1}{16\,580} - \frac{e}{0.842 \times 10^6} \right\} = 0$$

i.e.

$$e = 50.78 \text{ mm measured from the centroid towards } \underline{P}$$

## Example 6.7

Figure 6.14(a) shows the cross-section of a compound stanchion fabricated by attaching flange plates to two back-to-back channels. The characteristics of one channel are as indicated in Figure 6.14(b).

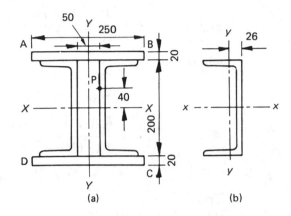

Figure 6.14

(a)  Calculate the stresses developed at the corners A, B, C and D of the section when it is subjected to a compressive load of 600 kN at point P.

(b)  Determine the maximum allowable eccentricities of this load from the centroid of the stanchion cross-section if no tensile stress is to be developed in the section. Make a sketch showing the boundary of the region within which the compressive force must be contained.

Data: For one channel: $A = 38 \text{ cm}^2$; $I_{xx} = 2490 \text{ cm}^4$; $I_{yy} = 264 \text{ cm}^4$.

(Nottingham Trent University)

## Solution 6.7

(a)

(1)  Calculate $I_{XX}$ and $I_{YY}$

$$I_{XX} = 171.13 \times 10^6 \text{ mm}^4$$
$$I_{YY} = 77.13 \times 10^6 \text{ mm}^4$$

**The calculation of $I_{XX}$ and $I_{YY}$ follows exactly the same procedure as illustrated in previous examples. The values for $I_{XX}$ and $I_{YY}$ given above should be checked by the reader before proceeding.**

$$Z_X = \frac{I_{XX}}{y_{max}} = \frac{171.13 \times 10^6}{120} = 1.426 \times 10^6 \text{ mm}^3$$

and

$$Z_Y = \frac{I_{YY}}{x_{max}} = \frac{77.13 \times 10^6}{125} = 0.617 \times 10^6 \text{ mm}^3$$

(2)   To determine the stresses

The stresses at the corners are given by

$$\sigma = \frac{P}{A} \pm \frac{P \times e_Y}{Z_X} \pm \frac{P \times e_X}{Z_Y} \qquad (1)$$

where

$$A = 2(38 \times 10^2) + 2(250 \times 20) = 17\,600 \text{ mm}^2$$

$$\therefore \sigma = \frac{600 \times 10^3}{17\,600} \pm \frac{600 \times 10^3 \times 40}{1.426 \times 10^6} \pm \frac{600 \times 10^3 \times 25}{0.617 \times 10^6}$$

$$= +34.09 \pm 16.83 \pm 24.31$$

Thus
stress at A $= +34.09 + 16.83 - 24.31 = +26.61 \text{ N/mm}^2$ (compressive)
stress at B $= +34.09 + 16.83 + 24.31 = +75.23 \text{ N/mm}^2$ (compressive)
stress at C $= +34.09 - 16.83 + 24.31 = + 41.57 \text{ N/mm}^2$ (compressive)
stress at D $= +34.09 - 16.83 - 24.31 = -7.05 \text{ N/mm}^2$ (tensile)

(b)   Equation (1) may be written

$$\sigma = \frac{600 \times 10^3}{17\,600} \pm \frac{600 \times 10^3 \times e_Y}{1.426 \times 10^6} \pm \frac{600 \times 10^3 \times e_X}{0.617 \times 10^6}$$

$$= 34.091 \pm 0.421 \times e_Y \pm 0.972 \times e_X$$

If the load is in the top right-hand quadrant of the section (i.e. in the quadrant containing corner B), then tension is most likely to occur at corner D, the stress at D being given by

$$\sigma_D = +34.091 - 0.421 \times e_Y - 0.972 \times e_X$$

If no tension is allowed to occur at D, then in the limit,

$$\sigma_D = +34.091 - 0.421 \times e_Y - 0.972 \times e_X = 0$$

Putting $e_X = 0$ gives

$$e_Y = 34.091/0.421 = 80.98 \text{ mm}$$

Putting $e_Y = 0$ gives

$$e_X = 34.091/0.972 = 35.07 \text{ mm}$$

The two points representing $e_X = 0$, $e_Y = 80.98$ mm and $e_X = 35.07$ mm, $e_Y = 0$ are marked as a and b on Figure 6.15. The line joining a to b defines

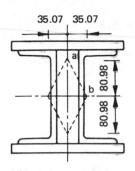

**Figure 6.15**

the boundary beyond which the load must not be placed if tension is not to develop at D.

Similar consideration of corners A, B and C will result in the core region of the section, as shown in Figure 6.15, within which the point of application of the load must lie to ensure that tension does not occur anywhere in the section.

## Example 6.8

Figure 6.16 shows a column loaded with three point loads. The positions of the loads are shown in both elevation and plan. The cross-section of the column is shown in the plan.

Determine the stresses at the four corners of the column due to the loading.

(University of Hertfordshire)

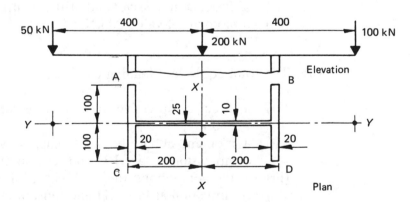

**Figure 6.16**

For reference purposes, the letters A, B, C and D have been added to the plan and the *X–X* and *Y–Y* axes have been identified.

153

*Solution 6.8*

(1)  Calculate $Z_X$ and $Z_Y$

$$I_{XX} = 327.94 \times 10^6 \text{ mm}^4$$
$$I_{YY} = 26.70 \times 10^6 \text{ mm}^4$$
$$Z_X = 1.640 \times 10^6 \text{ mm}^3$$
$$Z_Y = 0.267 \times 10^6 \text{ mm}^3$$

**Check these values before proceeding.**

(2)  Loads eccentric to the *Y–Y* axis

The 200 kN load is the only load eccentric to the *Y–Y* axis and gives rise to a moment about the *Y–Y* axis of value

$$M_Y = 200 \times (25 \times 10^{-3}) = 5 \text{ kNm}$$

The equivalent loading is thus
(i)   a direct compressive load at the centroid, of value 200 kN;
(ii)  a moment of value 5 kNm.

(3)  Loads eccentric to the *X–X* axis

**The two loads eccentric to the *X–X* axis give rise to moments of different sense, one clockwise and one anticlockwise about the *X–X* axis. The total effect is the algebraic sum of the two moments.**

The net moment about the *X–X* axis is given by

$$M_X = 100 \times (400 \times 10^{-3}) - 50 \times (400 \times 10^{-3}) = 20 \text{ kNm}$$

The equivalent loading is thus
(i)   a direct compressive load at the centroid, of value 150 kN;
(ii)  a moment of value 20 kNm.

(4)  The total equivalent loading is

(i)   a direct compressive load at the centroid, of value $P = 200 + 150 = 350$ kN;
(ii)  a moment about the *X–X* axis, of value $M_X = 20$ kNm, causing compression at face BD and tension at face AC;
(iii) a moment about the *Y–Y* axis, of value $M_Y = 5$ kNm, causing compression at face CD and tension at face AB.

(5)  The stresses at the corners of the section are given by

$$\sigma = \frac{P}{A} \pm \frac{M_X}{Z_X} \pm \frac{M_Y}{Z_Y}$$

where

$$A = 2(200 \times 20) + (360 \times 10) = 11.6 \times 10^3 \text{ mm}^2$$

$$\therefore \sigma = \frac{350 \times 10^3}{11.6 \times 10^3} \pm \frac{20 \times 10^6}{1.640 \times 10^6} \pm \frac{5 \times 10^6}{0.267 \times 10^6}$$

$$= 30.17 \pm 12.20 \pm 18.73$$

Thus,

$$\text{stress at A} = +30.17 - 12.20 - 18.73 = -0.76 \text{ (tensile)}$$
$$\text{at B} = +30.17 + 12.20 - 18.73 = +23.64 \text{ (compressive)}$$
$$\text{at C} = +30.17 - 12.20 + 18.73 = +36.70 \text{ (compressive)}$$
$$\text{at D} = +30.17 + 12.20 + 18.73 = +61.10 \text{ (compressive)}$$

## Example 6.9

A gravity dam, retaining water to a depth of 5 m, is shown in Figure 6.17. If the density of the water is 10 kN/m$^3$ and that of the dam material 20 kN/m$^3$, show on a drawing the stress distribution across the base.

(University of Hertfordshire)

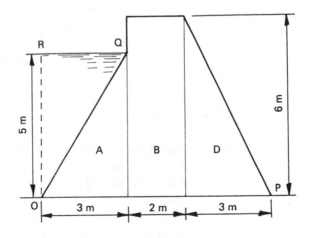

**Figure 6.17**

For reference purposes the parts of the dam have been lettered. It is necessary to locate the centre of gravity of the dam. This will coincide with the centroid of the cross-section, the position of which can be determined as before by taking moments of area.

It is normal practice to consider a unit length of this type of structure when calculating self-weight and water thrusts.

*Solution 6.9*

(1) To locate the centre of gravity of the dam

| Part | Area ($A$) (m$^2$) | distance $z$, measured from O (m) | $Az$ |
|------|------|------|------|
| A | 7.5 | 2.0 | 15.0 |
| B | 12.0 | 4.0 | 48.0 |
| D | 9.0 | 6.0 | 54.0 |
| | 28.5 | | 117.0 |

$$\bar{z} = \frac{\Sigma Az}{\Sigma A} = \frac{117.0}{28.5} = 4.105 \text{ m}$$

Thus, the weight of the dam acts at a distance 4.105 m to the right of O—that is, at an eccentricity $e = 4.105 - \frac{8}{2} = 0.105$ m from the centroid of the base (see Figure 6.18).

The self-weight of a 1 m length of the dam will be

$$28.5 \times 1 \times 20 = 570 \text{ kN}$$

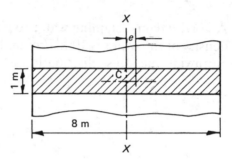

**Figure 6.18**

This is equivalent to

(i)   a direct compressive load through the centroid of the plan section, of value 570 kN;

(ii)  a moment, of value $570 \times 0.105 = 59.85$ kNm, causing compression at face P.

(2) Loading on the dam due to water pressure

**The total hydrostatic thrust on the sloping face of the dam will act perpendicular to the face and at two-thirds of the depth (i.e. at $\frac{5}{3}$ m above the base). A convenient solution is to determine the vertical and horizontal components of the total thrust.**

The value of the vertical component is given by the weight of the wedge-shaped body of water OQR. The horizontal component is given by multiplying the area of the face of the dam projected onto the vertical plane, by the average water pressure.

Vertical component of thrust = Area OQR $\times$ 1 $\times$ 10

$$= 5 \times \tfrac{3}{2} \times 10 = 75.0 \text{ kN}$$

This component acts through the centroid of the triangle OQR and at a distance of 1 m to the right of O, i.e. at an eccentricity of 3 m from the centroid of the base section. The vertical component of thrust is thus equivalent to

(i) a direct compressive load, of value 75.0 kN, through the centroid of the base;
(ii) a moment of value $75.0 \times 3.0 = 225.0$ kNm causing compression at face O.

Horizontal component of thrust $= \frac{1}{2}wh^2 = \frac{1}{2} \times 10 \times 5^2 = 125.0$ kN
This load acts at $\frac{5}{3}$ m above the base, providing a moment of value $125.0 \times \frac{5}{3} = 208.33$ kNm causing compression at face P.

**The horizontal component makes no contribution to the equivalent direct axial loading through the centroid of the base.**

(3) Total equivalent loading
The equivalent loading on the plan section is thus

(i) a direct compressive load of value $570.0 + 75.0 = 645.0$ kN;
(ii) a moment of value $59.85 - 225.00 + 208.33 = 43.18$ kNm, causing compression at face P.

**From now on the solution is straightforward and as in the previous examples.**

(4) To determine the stress distribution

The stresses at P and O are given by

$$\sigma = \frac{P}{A} \pm \frac{M_X}{Z_X}$$

where

$$A = 8 \times 1 = 8 \text{ m}^2$$

and

$$Z_X = bd^2/6 = 1 \times 8^2/6 = 10.667 \text{ m}^3$$

$$\therefore \sigma = \frac{645.0}{8} \pm \frac{43.18}{10.667}$$

$$= 80.63 \pm 4.05$$

Thus,

stress on base at P $= 80.63 + 4.05 = 84.68$ kN/m$^2$ (compressive) and

$$\text{at O} = 80.63 - 4.05 = 76.58 \text{ kN/m}^2 \text{ (compressive)}$$

The stress will vary linearly from O to P, as shown in Figure 6.19.

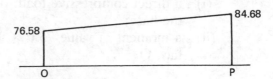

**Figure 6.19** Stress distribution diagram

## Example 6.10

A solid 7 m high vertical telegraph pole PQ, 250 mm in diameter at the ground, carries at its top two groups of horizontal wires A and B. In plan view these wires make an angle of 120° (Figures 6.20a and b). The total tension in group A is 1400 N and in group B 2300 N. The pole is restrained by a cable PR that makes an angle of 30° to the pole and, in plan view, makes an angle of 90° to the group A wires. The tension in this cable is 1800 N.

Determine the maximum tensile stress in the pole at the base Q.

(*Note* $I_{XX}$ for a solid circular section about a diameter is $\pi d^4/64$, where $d$ is the diameter.)

(Manchester University)

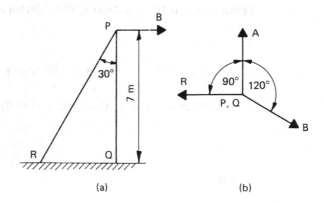

**Figure 6.20**

## Solution 6.10

(1) The tension in cable PR can be resolved vertically and horizontally:

$$\text{vertical component} = 1800 \cos30° = 1558.85 \text{ N}$$

This vertical component acts down the longitudinal axis of the pole and results in a vertical compressive force of 1558.85 N through the centroid of the cross-section of the pole.

$$\text{Horizontal component} = 1800 \sin30° = 900 \text{ N}$$

158

(2)   Consider the plan view at the top of the pole (Figure 6.21). The tensile forces in wires A and B are partially balanced by the horizontal component of the force in the cable PR. The out-of-balance force must be resisted by the development of bending stresses in the pole. The out-of-balance force can be determined by resolving the three known forces in the *x* and *y* direction and summing the components.

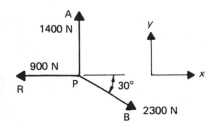

Figure 6.21

Resolving the horizontal forces at P:

| Force | | $y$ component | $x$ component |
|---|---|---|---|
| Wires A | 1400 N | +1400.00 | 0.00 |
| Wires B | 2300 N | −1150.00 | +1991.86 |
| Cable | 900 N | 0.00 | −900.00 |
| | | +250.00 | +1091.86 |

∴ Resultant horizontal force bending pole = $(250^2 + 1091.86^2)^{1/2}$ = 1120.12 N

(3)   The equivalent loading on the base of the pole is thus

   (i)   a direct compressive force, of value 1558.86 N, through the centroid;
   (ii)   a moment about a diameter, of value 1120.12 × 7 = 7840.84 Nm.

(4)   The maximum tensile stress in the cross-section at Q will be

$$\sigma = \frac{P}{A} - \frac{M}{Z}$$

where

$$A = \pi \times d^2/4 = \pi \times 250^2/4 = 49.09 \times 10^3 \text{ mm}^2$$

and

$$Z = \frac{I}{y_{max}} = \frac{\pi \times d^4/64}{d/2} = \frac{\pi \times d^3}{32} = \frac{\pi \times 250^3}{32} = 1.534 \times 10^6 \text{ mm}^3$$

$$\therefore \sigma = \frac{1558.85}{49.09 \times 10^3} - \frac{7840.84 \times 10^3}{1.534 \times 10^6}$$

$$= 0.032 - 5.111 = -5.08 \text{ N/mm}^2$$

$$\therefore \text{maximum tensile stress} = \underline{5.08 \text{ N/mm}^2}$$

**Example 6.11**

Figure 6.22 shows the section of a symmetrical prestressed concrete beam in which the eccentricity '$e$' of the tendons is 105 mm. The cross-section area is $77.6 \times 10^3$ mm$^2$ and $I_{XX}$ is $1.6 \times 10^9$ mm$^4$.

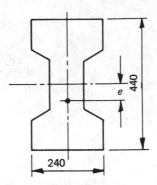

**Figure 6.22**

(a) Calculate the maximum allowable prestressing force if, at the prestressing stage, the allowable stresses are 1 N/mm$^2$ tension and 20 N/mm$^2$ compression.

(b) What applied moment can then be resisted if the allowable stresses under load are zero tension and 20 N/mm$^2$ compression?

(University of Hertfordshire)

**The prestressing force in the tendon develops compressive bending stresses in the lower part of the cross-section of the beam and tensile bending stresses in the top part.**

**The applied bending moment to be carried by the beam will be such as to develop bending stresses of opposite sense to those developed by the prestressing tendon.**

*Solution 6.11*

(a)

$$\text{Elastic section modulus } Z_X = \frac{I_{XX}}{y_{\text{max}}} = \frac{1.6 \times 10^9}{220} = 7.27 \times 10^6 \text{ mm}^3$$

If the allowable prestressing force is $P$ kN, then the moment due to the prestressing force is given by

$$M_X = P \times e = (P \times 10^3) \times 105 \text{ Nmm}$$

160

The maximum compressive stress occurs at the bottom face and is given by

$$\sigma = \frac{P}{A} + \frac{M_X}{Z_X}$$

$$= \frac{(P \times 10^3)}{77.6 \times 10^3} + \frac{(P \times 10^3) \times 105}{7.27 \times 10^6}$$

$$= P(12.887 + 14.443) \times 10^{-3}$$
$$= 27.330P \times 10^{-3} \text{ N/mm}^2$$

But the maximum compressive stress is limited to 20 N/mm². Hence,

$$27.330P \times 10^{-3} \text{ N/mm}^2 < 20$$

$$\underline{P < 731.80 \text{ kN}}$$

The maximum tensile stress occurs at the top face and is given by

$$\sigma = \frac{P}{A} - \frac{M_X}{Z_X}$$

$$= \frac{(P \times 10^3)}{77.6 \times 10^3} - \frac{(P \times 10^3) \times 105}{7.27 \times 10^6}$$

$$= P(12.887 - 14.443) \times 10^{-3}$$
$$= -1.556P \times 10^{-3} \text{ N/mm}^2$$

But the maximum tensile stress is limited to 1 N/mm². Hence,

$$1.556P \times 10^{-3} \text{ N/mm}^2 < 1$$

i.e.

$$\underline{P < 642.67 \text{ kN}}$$

The maximum allowable value for $P$ is therefore the lower of these two figures—that is, 642.67 kN. With this value for $P$, the compressive stress at the lower face will be

$$\frac{642.67}{731.80} \times 20 = 17.56 \text{ N/mm}^2$$

(b)

**The allowable stresses are 20 N/mm² compressive and zero tensile. The top face has an initial tensile prestressing stress of 1 N/mm², which, when the beam is subject to an applied moment, may be changed to a compressive stress of 20 N/mm². Thus, an applied bending stress of 21 N/mm² could be accommodated in the top face. The lower face has an initial compressive prestressing stress of 17.56 N/mm², which may be changed to zero N/mm². Thus, an applied bending stress of 17.56 N/mm² only may be accommodated in the lower face. Thus, the maximum external applied bending moment which can safely be carried is limited by the corresponding bending stress of 17.56 N/mm².**

If the allowable applied bending moment is $M_X$, the corresponding bending stress is given by

$$\sigma = \frac{M_X}{Z_X} = \frac{M_X}{7.27 \times 10^6}$$

But $\sigma$ is limited to 17.56 N/mm². Therefore,

$$M_X = 17.56 \times 7.27 \times 10^6 \times 10^{-6} = \underline{127.66 \text{ kNm}}$$

## 6.5 Problems

**6.1** The bracket shown in Figure P6.1(a) carries an inclined point load of 5 kN at A. The cross-section of the bracket at the line BC is shown in Figure P6.1(b). Determine the maximum tensile stress at the section BC and the distance of the neutral axis of the same section from its centre of area.

(Manchester University)

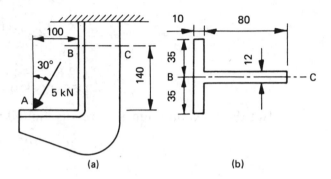

**Figure P6.1**

**6.2** Figure P6.2 shows the cross-section through a brickwork chimney stack. The stack is rigidly built into the ground and is 18 m high.

The density of the brickwork is 2300 kg/m³ and the most adverse wind loading causes simultaneous pressures of 0.5 kN/m² and 0.3 kN/m² on the longer and shorter sides, respectively.

The design criteria are that the compressive stress developed in the brickwork is not to exceed 2.0 N/m² and no tension is to be developed. Ascertain whether these conditions are satisfied by the design.

(Nottingham Trent University)

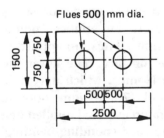

**Figure P6.2**

**6.3** Calculate the maximum eccentricity at which a concentrated point load can be placed along the $X$ and $Y$ axes before tension develops in the section shown in Figure P6.3

(Liverpool University)

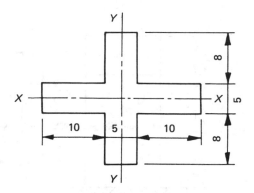

**Figure P6.3**

**6.4** A hollow cast-iron column with the cross-section shown in Figure P6.4 is subjected to a vertical load of 80 kN which acts through corner A. Calculate the maximum compressive and tensile stresses in the column. Determine also the minimum vertical load which must be applied at corner B in order that there is no resultant tensile stress caused by the two loads.

(Liverpool University)

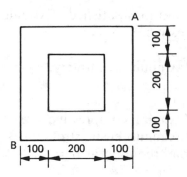

**Figure P6.4**

**6.5** A check is to be made on a structure supporting an oil supply tank. The tank is mounted on a saddle which transfers the load uniformly to a hollow masonry pillar 9.0 m high and of overall cross-sectional dimensions 6.0 m by 1.5 m, the wall thickness being 0.4 m. The overall height to the top of the tank is 12.0 m, which extends the full length of the structure as indicated in Figure P6.5(a). The cross-section of the masonry is shown in Figure P6.5(b).

The capacity of the tank is 40 000 litres and, when empty, the tank plus the saddle weigh 8150 kg. The density of the oil is 800 kg/m$^3$ and of the masonry is 1800 kg/m$^3$.

The maximum pressure exerted by the wind when blowing perpendicular to one face may be taken as 0.8 kN/m² over the total exposed area.

Determine whether or not the structure satisfies the criteria that the maximum permitted compressive stress is 850 kN/m² and tension is not allowed.

(Sheffield University)

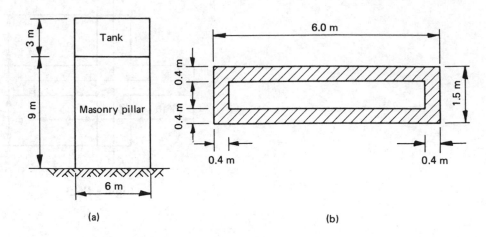

**Figure P6.5**

## 6.6 Answers to Problems

**6.1**  7.65 N/mm²; 37.28 mm

**6.2**  Conditions satisfied. Extreme stresses are 0.67 N/mm² and 0.14 N/mm² compressive.

**6.3**  Maximum eccentricity, $e_X = \pm 2.61$ mm
$\qquad\qquad\qquad\qquad e_Y = \pm 1.89$ mm

**6.4**  Maximum compressive stress, at A = 3.87 N/mm²
Maximum tensile stress, at B = 2.53 N/mm²
Load required at B = 52.34 kN

**6.5**  Conditions satisfied. The stresses at the outer long faces of the pillar are:

(i)  when the tank is full, 400.83 kN/m² and 63.97 kN/m² (compressive);
(ii) when the tank is empty, 342.27 kN/m² and 5.41 kN/m² (compressive).

# 7  Shear Stress

## 7.1  Contents

Shear stress • Complementary shear stress • Shear in rectangular beam sections • Shear in non-rectangular beam sections • Shear stress distribution in beams • Compound beams • Shear in bolted and welded connections.

This chapter is mainly concerned with the distribution of shear stress in simple and compound beam sections, together with the calculation of the shearing forces in the bolted or welded connections of compound beams.

## 7.2  The Fact Sheet

### (a)  Average Shear Stress

The average shear stress acting over a surface on which a shearing force is acting is given by

$$\text{average shear stress } (\tau) = \frac{\text{shearing force}}{\text{area over which shearing force acts}} .$$

### (b)  Shear Strain

The action of shear stress on a material gives rise to *shear strain*. Shear strain $(\gamma)$ is measured as the angular distortion of a small rectangular element of the material. Shear strain is measured in radians and is dimensionless.

### (c)  Shear Modulus

For a linearly elastic material shear stress and shear strain are related by the *shear modulus* $(G)$, sometimes referred to as the modulus of rigidity. The shear modulus is given by the formula

$$\text{shear modulus } (G) = \frac{\text{shear stress } (\tau)}{\text{shear strain } (\gamma)}$$

165

### (d)  Complementary Shear Stress

A shear stress acting on either of two planes which intersect at right angles is always accompanied by a *complementary* shear stress of equal magnitude but of opposite sign acting on the other plane.

### (e)  Shear Stresses in Beam Sections

The general formula for calculating the shear stress in a beam section which is subjected to a shearing force is given by

$$\tau = \frac{QA\bar{y}}{Ib}$$

where  $\tau$ = the shear stress;
$\quad Q$ = the transverse shear force acting at the section;
$\quad A$ = the area to the outside of the level where shear stress is being calculated;
$\quad \bar{y}$ = the distance to the centroid of the area $A$ measured from the neutral axis;
$\quad I$ = the second moment of area of the *whole* cross-section about the neutral axis; and
$\quad b$ = the width of the cross-section at the level where shear stress is being calculated.

The terms used in this equation are illustrated in Figure 7.1.

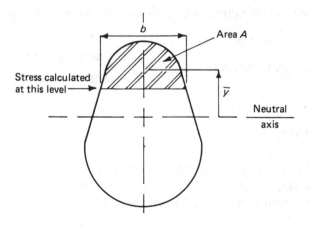

**Figure 7.1**

In commonly occurring structural sections the maximum shear stress calculated by use of this formula usually occurs at the level of the neutral axis. In rectangular, T-shaped and I-shaped beams and other commonly occurring sections the shear stress varies parabolically throughout the depth of the section, with abrupt changes of stress where the geometry of the section changes suddenly, such as where the web and flanges of an I section meet.

166

### (f)  Shear Stresses in Rectangular Beam Sections

For a rectangular beam section the *maximum* shear stress is 50% greater than the *average* shear stress.

## 7.3  Symbols, Units and Sign Conventions

$A$ = the area of cross-section of a beam to the outside of the level where shear stress is being calculated ($mm^2$)

$b$ = the width of the beam cross-section at the level where shear stress is being calculated (mm)

$d$ = bolt diameter (mm) *or* depth of section (mm)

$h$ = the perpendicular distance between an axis passing through the centroid of a section and a parallel axis (mm)

$I$ = the second moment of area of the *whole* cross-section about the neutral axis ($mm^4$)

$I_{XX}$ = the second moment of area of a section about an $X–X$ axis ($mm^4$)

$I_{CC}$ = the second moment of area of a section about an axis passing through the centroid of the section ($mm^4$)

$M$ = bending moment (kNm)

$P$ = force (kN)

$Q$ = the shearing force acting across the beam section (kN)

$y$ = the distance from the neutral axis to the level at which the *bending stress* is calculated in the formula ($\sigma = yM/I$) (mm)

$\bar{y}$ = the distance to the centroid of the area $A$ measured from the neutral axis, as used in the shear stress formula (mm)

$z$ = a distance used for the determination of the position of the neutral axis of a section ($y$ is more commonly used for this purpose but $z$ is used in this text to avoid confusion with the terms $y$ and $\bar{y}$ as defined above) (mm)

$\tau$ = shear stress ($N/mm^2$)

## 7.4  Worked Examples

### Example 7.1

Figure 7.2 shows a cross-section through a reinforced concrete footing which supports a centrally placed column. In plan the footing measures 2.5 × 2.5 m and, because of the loading transmitted down the column, there is a uniform pressure under the footing of 200 $kN/m^2$. Calculate the average shear stress at section $X–X$ and, if the allowable shear stress is 0.5 $N/mm^2$, state whether the average shear stress is acceptable.

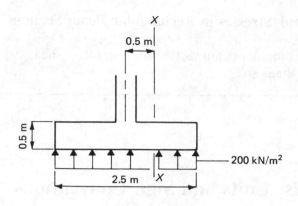

**Figure 7.2**

*Solution 7.1*

**This is a fairly straightforward problem as an introduction to this chapter. Remembering that the shearing force acting on any section is the sum of all forces to one side of the section:**

$$\text{area of base to right of } X\text{-}X = \text{width of base} \times \text{length to right of } X\text{-}X$$
$$= 2.5 \times (2.5/2 - 0.5)$$
$$= 1.875 \text{ m}^2$$

$$\text{Shearing force at section } X\text{-}X = \text{pressure under base} \times \text{area of base to right of } X\text{-}X$$

$$\therefore \text{Shearing force at section } X\text{-}X = 200 \times 1.875 = 375 \text{ kN}$$

$$\text{Average shear stress at } X\text{-}X = \frac{\text{shearing force at section}}{\text{area of cross-section}}$$

$$= \frac{375 \times 10^3}{2500 \times 500}$$

$$= \underline{0.3 \text{ N/mm}^2}$$

As this shear stress is less than the allowable figure of 0.5 N/mm², this shear stress is acceptable.

## Example 7.2

A pressure vessel with 750 mm internal diameter is to be formed by connecting two semicylindrical sections as shown in Figure 7.3. The connections are to be made with 25 mm diameter rivets, for which the maximum permissible average stress is 125 N/mm², equally spaced along the seams. If the internal pressure is to be 1.5 N/mm², what should be the pitch of the rivets?

(Birmingham University)

**The forces to be resisted by the rivets could be determined by use of standard formulae for hoop stresses. In this example, however, the necessary formulae will**

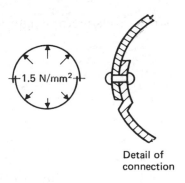

Detail of
connection

**Figure 7.3**

be developed from first principles, the calculations being based on a unit length of cylinder of 1 m. Refer to Figure 7.4, which shows the forces acting on one half of the vessel and which are in equilibrium with the internal forces due to the internal pressure.

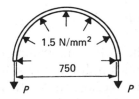

**Figure 7.4**

### Solution 7.2

Let the force in each side of the wall = $P$. Then, equating internal and external forces,

$$2 \times P = \text{internal pressure} \times \text{projected area over which pressure acts}$$
$$= \text{internal pressure} \times \text{length} \times \text{diameter}$$

$$\therefore P = \frac{1.5 \times 10^3 \times (1 \times 0.75)}{2}$$

$$= 562.5 \text{ kN/m}$$

Note that in this last calculation all the terms have been converted to units of kN and metres. The force $P$ must be developed by shearing action in the rivets.

Permissible average shear stress in rivets ($\tau$) = 125 N/mm$^2$

$\therefore$ Permissible shearing force per rivet = $\frac{1}{4}\pi d^2 \times \tau = \frac{1}{4}\pi \times 25^2 \times 125 \times 10^{-3}$
$$= 61.36 \text{ kN}$$

Number of rivets/metre length to resist shear on each side of the vessel

$$= \frac{\text{total force } (P)}{\text{shear force/rivet}}$$

$$= \frac{562.5}{61.36}$$

$$= 9.17 \text{ rivets/m}$$

$$\text{Pitch of rivets} = \frac{\text{length of cylinder}}{\text{No. of rivets}}$$

$$= \frac{1000}{9.17}$$

$$= 109.05 \text{ mm}$$

**Sensibly this answer should be rounded off to 100 mm if this were a real design problem.**

### Example 7.3

(a) Show that in a beam of rectangular cross-section, subjected to a transverse shearing force, the value of the maximum shear stress is 50% greater than the average shear stress. Sketch the shear stress distribution.

(b) The I-section beam shown in Figure 7.5 is subjected to a transverse shearing force of 500 kN.

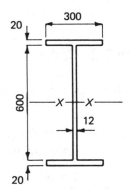

**Figure 7.5**

(i) Obtain the value of the maximum shear stress and sketch the distribution of shear stress in the web.

(ii) Calculate the percentage difference between the maximum and the average shear stress in the web of this section. Comment on the practical significance of this result in respect of design calculations.

(University of Portsmouth)

170

*Solution 7.3*

(a)
**This is a standard proof which emphasises a feature of design of rectangular sections which is utilised in the design of, for example, beams and joists in timber construction.**

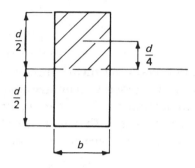

**Figure 7.6**

Let the dimensions of the beam be as shown in Figure 7.6. The maximum shear stress will occur at the level of the neutral axis, which is at mid-depth of the section. The shear stress, $\tau$, is given by the standard formula

$$\tau = \frac{QA\bar{y}}{Ib}$$

where, for a rectangle,

$$I = \frac{bd^3}{12}$$

and, for shear at mid-depth,

$$A\bar{y} = b \times (d/2) \times (d/4) = bd^2/8$$

Hence,

$$\tau_{max} = \frac{Q \times (bd^2/8)}{(bd^3/12) \times b}$$

$$= \frac{3}{2}\frac{Q}{bd}$$

The *average* shear stress is given by the total shearing force ($Q$) divided by the gross beam section ($b \times d$). Hence, from the above equation,

$$\tau_{max} = 1.5 \times \text{average shear stress}$$
$$= 50\% \text{ greater than the average}$$

(b)
**This beam section is symmetrical and, hence, the neutral axis passes through the centroid at mid-depth, which is also the position where the maximum shear stress will occur. It is first necessary to calculate the second moment of area about the**

171

neutral axis. In this case this is best done by calculating the second moment of area of the gross rectangular section 300 mm wide × 640 mm deep and subtracting from it the second moment of area of the rectangle which is removed to form the I-section.

Hence,

$$I = \frac{1}{12}\,300 \times 640^3 - \frac{1}{12}\,288 \times 600^3$$

$$= 1369.60 \times 10^6 \text{ mm}^4$$

The maximum shear stress occurs at the level of the neutral axis and the term '$A\bar{y}$' in the general shear stress equation represents the first moment of area of the beam section to either side of the neutral axis. Considering the area above the neutral axis, this first moment of area is the sum of the moments of area of the flange and that part of the web above the neutral axis.

$$\therefore A\bar{y} = Ay_{\text{flange}} + Ay_{\text{web}}$$
$$= (300 \times 20 \times 310) + (12 \times 300 \times 150)$$
$$= 2.40 \times 10^6 \text{ mm}^3$$

Hence,

$$\tau_{\text{max}} = \frac{QA\bar{y}}{Ib}$$

$$= \frac{500 \times 10^3 \times 2.40 \times 10^6}{1369.60 \times 10^6 \times 12}$$

$$= 73.01 \text{ N/mm}^2$$

The question asks for a sketch of the shear stress distribution in the web. The student is expected to know that the shear stress varies parabolically throughout the depth of the web, with maximum stress occurring at the level of the neutral axis, as shown in Figure 7.7. Most standard textbooks will demonstrate the proof that the shear stress distribution is in fact parabolic.

In the calculation of *average* shear stress in the web the depth of the web can be taken as the overall beam depth and the stress in the flange neglected, as is commonly done in steel design.

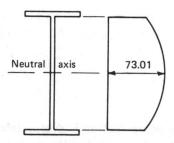

**Figure 7.7**  Distribution of shear stress in web

The *average* shear stress is given by

$$\tau_{av} = \frac{Q}{b \times d}$$

$$= \frac{500 \times 10^3}{12 \times 640}$$

$$= 65.10 \text{ N/mm}^2$$

$$\% \text{ difference} = \frac{73.01 - 65.10}{65.10} = 12.15\%$$

The percentage difference between the maximum and average shear stresses is small and, hence, in the design of steel I-sections it is justified to base design calculations on the more easily calculated average value.

## Example 7.4

For the section shown in Figure 7.8 plot the distribution of shear stress along the vertical axis and determine the maximum shear force which can be carried by the section if the permissible maximum shear stress must not exceed 90 N/mm².

(Liverpool University)

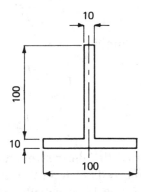

**Figure 7.8**

## *Solution 7.4*

**This example differs from the previous one in that it is not a symmetrical section and the position of the neutral axis, which is also the centroid of the cross-section, must first be determined.**

**To determine the centroid, split the section into rectangular components and take moments of area about the base of the section. The parallel axis theorem can then be used to calculate the second moment of area of the section about the neutral axis. This is best set out in tabular form, as follows:**

173

| Part | Area $(A)$ $(mm^2)$ | $z$ $(mm)$ | $Az$ $(mm^3 \times 10^3)$ | $I_{CC}(bd^3/12)$ $(mm^4 \times 10^3)$ | $h = (z - \bar{z})$ $(mm)$ | $Ah^2$ $(mm^4 \times 10^3)$ |
|---|---|---|---|---|---|---|
| A | $100 \times 10$ $= 1000$ | 5 | 5 | 8.33 | $32.5 - 5.0$ $= 27.5$ | 756.25 |
| B | 1000 | 60 | 60 | 833.33 | 27.5 | 756.25 |
| | 2000 | | 65 | 841.66 | | 1512.50 |

$$\bar{z} = \frac{\Sigma Az}{\Sigma A} = \frac{65 \times 10^3}{2000} = 32.5 \text{ mm}$$

$$\begin{array}{c} 841.66 \times 10^3 \\ 1512.50 \times 10^3 \end{array}$$

$$I_{XX} = 2354.16 \times 10^3 \text{ mm}^4$$

The maximum shear stress occurs at the neutral axis and in this case it is easier to calculate the $A\bar{y}$ term in the shear stress formula by considering the area *above* the neutral axis, as there is only one rectangular section to this side of the neutral axis.

Hence, considering the stress at the neutral axis, which has a limiting permissible value of 90 N/mm$^2$,

$$\tau = \frac{QA\bar{y}}{Ib}$$

or, rearranging,

$$Q = \frac{\tau Ib}{A\bar{y}} = \frac{90 \times 2354.16 \times 10^3 \times 10 \times 10^{-3}}{(10 \times 77.5) \times 77.5/2}$$

$$= \underline{70.55 \text{ kN}}$$

To draw the shear stress diagram, it is necessary to calculate the stresses at the levels marked (1) and (2) in Figure 7.9. Note that in the following calculations the term $A\bar{y}$ has been calculated for the area of the section *below* the level being considered, shown hatched in Figure 7.9.

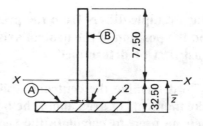

**Figure 7.9**

Thus, at level (1),

$$\tau = \frac{QA\bar{y}}{Ib}$$

$$= \frac{(70.55 \times 10^3) \times (\{100 \times 10\} \times \{32.5 - 5\})}{2354.16 \times 10^3 \times 10}$$

$$= \underline{82.41 \text{ N/mm}^2}$$

And, at level (2),

$$\tau = \frac{QA\bar{y}}{Ib}$$

$$= \frac{(70.55 \times 10^3) \times (\{100 \times 10\} \times \{32.5 - 5\})}{2354.16 \times 10^3 \times 100}$$

$$= \underline{8.24 \text{ N/mm}^2}$$

Note that this last answer could have been obtained by simply multiplying the stress at level (1) by the ratio between the width of the web and the width of the flange.

The shear stress diagram is as shown in Figure 7.10. Again the student should know that the shear stress varies parabolically throughout the depth of the section, with zero stress at the extreme top and bottom fibres of the section and an abrupt jump in the shape of the diagram at the junction of the web and flange.

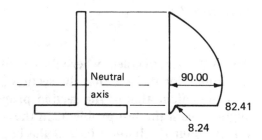

**Figure 7.10**

## Example 7.5

A universal beam section is strengthened by welding a 200 mm × 20 mm plate to the top flange as shown in Figure 7.11.

(a)  For the compound section determine:

    (i)   the depth ($\bar{z}$) to the neutral axis;
    (ii)  the second moment of area about the neutral axis.

(b)  If this compound section is then used as a simply supported beam and the permissible bending stresses are limited to 140 N/mm² and 80 N/mm² in tension and compression, respectively, and the permissible shear stress is 70 N/mm², then determine:

(i) the maximum bending moment;
(ii) the maximum shearing force which the beam can sustain.

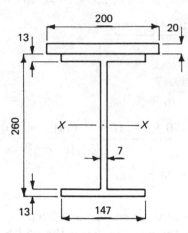

For the universal beam:
area = 54.60 cm²;
$I_{XX}$ = 6582 cm⁴.

**Figure 7.11**

(Coventry University)

### *Solution 7.5*

(a)
**Again this is a problem where the calculation of the section properties is best done in tabular form. Note that the section properties of the universal beam are given and the calculation of the section properties of the composite section will differ slightly from the last problem. In the following calculations moments of area are taken about the bottom face of the beam.**

| Part | Area ($A$) (mm²) | $z$ (mm) | $Az$ (mm³ × 10³) | $I_{CC}$ (mm⁴ × 10⁶) | $h = (z - \bar{z})$ (mm) | $Ah^2$ (mm⁴ × 10⁶) |
|---|---|---|---|---|---|---|
| Plate | 200 × 20 = 4000 | 270 | 1080.0 | 0.13 | 270 − 189.2 = 80.8 | 26.11 |
| I-beam | 5460 | 130 | 709.8 | 65.82 | 59.2 | 19.14 |
| | 9460 | | 1789.8 | 65.95 | | 45.25 |

$$\bar{z} = \frac{\Sigma Az}{\Sigma A} = \frac{1789.8 \times 10^3}{9460} = \underline{189.2 \text{ mm}}$$

$$\begin{array}{r} 65.95 \times 10^6 \\ 45.25 \times 10^6 \end{array}$$

$$I_{XX} = \underline{111.20 \times 10^6 \text{ mm}^4}$$

(b)

The bending moment that will cause limiting tensile stress on the bottom face of the beam is given by

$$M = \frac{\sigma_{\text{ten}}I}{y} = \frac{140 \times 111.20 \times 10^6 \times 10^{-6}}{189.20} = 82.28 \text{ kNm}$$

The bending moment that will cause limiting compressive stress on the top surface of the steel plate is given by

$$M = \frac{\sigma_{\text{com}}I}{y} = \frac{80 \times 111.20 \times 10^6 \times 10^{-6}}{(260 + 20 - 189.20)} = 97.97 \text{ kNm}$$

**Note that, as the plate and beam are assumed to bend together, the maximum compressive stress occurs at the top of the steel plate and not at the top of the universal beam.**

**The maximum moment that the beam can carry is given by the lower of these two figures, because if the lower figure is exceeded, then the beam will be overstressed in tension.**

Hence, the maximum bending moment the beam can sustain = <u>82.28 kNm</u>.

The maximum shear stress occurs at the level of the neutral axis and, taking moments of area of that part of the beam below the neutral axis, the shear force that will give the limiting shear stress of 70 N/mm² is given by

$$\tau = \frac{QA\bar{y}}{Ib}$$

or

$$Q = \frac{\tau Ib}{A\bar{y}}$$

where

$$\begin{aligned}
A\bar{y} &= Ay_{\text{flange}} + Ay_{\text{web}} \\
&= (147 \times 13 \times \{189.20 - 6.5\}) + (7 \times \{189.20 - 13\}^2/2) \\
&= 457.80 \times 10^3 \text{ mm}^3 \\
\therefore Q &= \frac{70 \times 111.20 \times 10^6 \times 7 \times 10^{-3}}{457.80 \times 10^3} \\
&= \underline{119.02 \text{ kN}}
\end{aligned}$$

**You could have obtained the same answer by considering that part of the beam above the neutral axis but in that case you would have to consider three rectangular components of area in your answer.**

### Example 7.6

Figure 7.12 shows the cross-section of a steel beam fabricated by welding a 180 mm wide by 12 mm thick plate to the top flange of a doubly symmetric

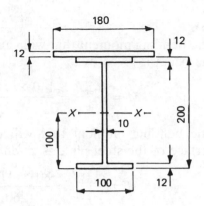

**Figure 7.12**

200 mm × 100 mm I-section. The cross-sectional area of the I-beam is 4160 mm² and its second moment of area $I_{XX}$ is $25.78 \times 10^6$ mm⁴. It is subjected to a shearing force of 50 kN.

(a) Determine and sketch the distribution of vertical shear stress over the depth of the section, showing all principal values.
(b) Compare the maximum shear stress in the web from the analysis in (a) with the average shear stress based on the assumption that the total shear load is carried uniformly by the web alone.
(c) Determine the load per unit length of the beam that must be resisted by the welded connection.

(Nottingham Trent University)

### Solution 7.6

(a)

**In this example the student is required to draw the shear stress diagram, marking all principal values. This means that the shear stress equation must be repeatedly applied at each level in the beam section where there is a change in geometry and at the level of the neutral axis. The calculations can be set out as in the previous example.**

| Part | Area $(A)$ (mm²) | $z$ (mm) | $Az$ (mm³ × 10³) | $I_{CC}$ (mm⁴ × 10⁶) | $h = (z - \bar{z})$ (mm) | $Ah^2$ (mm⁴ × 10⁶) |
|---|---|---|---|---|---|---|
| Plate | 2160 | 206 | 444.96 | 0.03 | 69.77 | 10.51 |
| I-beam | 4160 | 100 | 416.00 | 25.78 | 36.23 | 5.46 |
| | 6320 | | 860.96 | 25.81 | | 15.97 |

$$\bar{z} = \frac{\Sigma Az}{\Sigma A} = \frac{860.96 \times 10^3}{6320} = 136.23$$

$$\begin{array}{r} 25.81 \times 10^6 \\ 15.97 \times 10^6 \end{array}$$

$$\underline{I_{XX} = 41.78 \times 10^6 \text{ mm}^4}$$

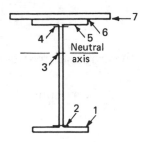

**Figure 7.13**

The stresses at the levels marked 1–7 in Figure 7.13 can now be calculated using the general shear stress equation. Remember that the term $A\bar{y}$ is taken as the first moment of area of that area to the outside of the level being considered, irrespective of whether the level is above or below the neutral axis. As this requires the repetitive use of a single equation, this is again best tabulated.

$$\tau = \frac{QA\bar{y}}{Ib} = \frac{50 \times 10^3}{41.78 \times 10^6} \frac{A\bar{y}}{b} = 1.20 \times 10^{-3} \frac{A\bar{y}}{b}$$

Hence,

| At level | $b$ | $A\bar{y}$ | $\tau = 1.20 \times 10^{-3} \times A\bar{y}/b$ (N/mm²) |
|---|---|---|---|
| 1 | 100 | $100 \times 12 \times (136.23 - 6)$ $= 156.28 \times 10^3$ | $1.20 \times 10^{-3} \times 156.28 \times 10^3/100$ $= 1.88$ |
| 2 | 10 | $156.28 \times 10^3$ | 18.75 |
| 3 | 10 | $233.44 \times 10^3$ | 28.01 |
| 4 | 10 | $220.03 \times 10^3$ | 26.40 |
| 5 | 100 | $220.03 \times 10^3$ | 2.64 |
| 6 | 100 | $150.70 \times 10^3$ | 1.81 |
| 7 | 180 | $150.70 \times 10^3$ | 1.00 |

Hence, the shear stress distribution diagram is as shown in Figure 7.14.

Again the student should realise that for this shape of cross-section the general shape of the stress diagram will be parabolic, with abrupt changes of stress at the intersection of the web and flanges and where the top flange and flange plate meet.

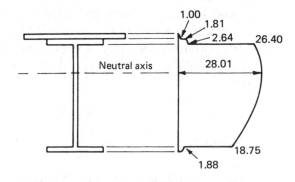

**Figure 7.14**

179

(b)

$$\text{Average shear stress in web} = \frac{Q}{bd} = \frac{50 \times 10^3}{10 \times 200} = 25.0 \text{ N/mm}^2$$

$$\text{Maximum shear stress} = 28.01 \text{ N/mm}^2$$

Hence,

$$\% \text{ difference} = \frac{28.01 - 25.0}{25.0} \times 100 = 12.04\%$$

(c)

Shear stress in beam at junction with plate $= \tau = 1.81 \text{ N/mm}^2$

**This is the *vertical* shear stress at this level, but there will also be a horizontal *complementary* shear stress of equal magnitude at this level which will give rise to a horizontal shear force which must be transmitted between the two structural components by the welded connection.**

Considering a unit length of beam of 1 m:

Shear *force* transmitted between beam and plate $= \tau \times$ area of shear surface
$$= 1.81 \times (1000 \times 100) \times 10^{-3}$$
$$= \underline{181.0 \text{ kN/m}}$$

$\therefore$ Load per unit length to be resisted by the welds $= \underline{181.0 \text{ kN}}$

**Note that in the calculation of this last answer the shear stress in the *beam* has been multiplied by the *horizontal* plan area of the beam at the beam–plate interface to give the shear force to be resisted by the welds. The same answer could have been obtained by using the stress in the *plate* and multiplying this stress by the plan area of the plate.**

## Example 7.7

A timber beam is simply supported at both ends and carries a central point load of 50 kN. The beam is fabricated from two 250 mm × 250 mm timbers which are bolted together as shown in Figure 7.15. The timbers are also keyed together by wooden keys which are spaced 700 mm apart. The maximum allowable average shear stress in the keys is 2 N/mm² and it is assumed that the keys resist all the longitudinal shear between the main timbers.

Calculate the least required thickness of the keys, shown as dimension $t$ in the figure.

(Coventry University)

*Solution 7.7*

**The assumption in this question is that the keys are carrying all the shear *force* along the horizontal plane between the main timbers, ensuring that when the**

180

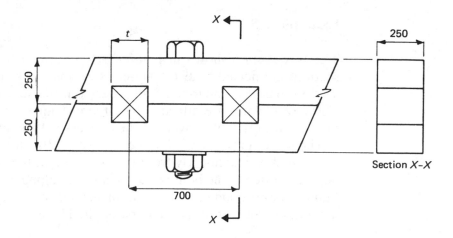

**Figure 7.15**

beam bends the two main timbers act together as a compound beam. It is thus necessary to calculate the shear *force* along the horizontal interface between the two timbers.

For a centrally placed point load the magnitude of the vertical shear force, $Q$, will be constant along the span and will equal one-half the magnitude of the point load. If in doubt, draw the shear force diagram.

Hence,

$$Q = \tfrac{1}{2} \times 50 = 25 \text{ kN}$$

$$\textit{Average shear stress across section} = \frac{Q}{bd} = \frac{25\,000}{250 \times 500} = 0.2 \text{ N/mm}^2$$

*Maximum* shear stress in a rectangular section $= 1.5 \times$ average stress
$$= 1.5 \times 0.2 = 0.3 \text{ N/mm}^2$$

This maximum shear stress occurs at the level of the neutral axis and there will be a *complementary* shear stress of equal magnitude acting along the horizontal plane between the two main timbers. This will give rise to a shearing *force* along this plane, which is resisted by the keys.

Horizontal shear stress $= 0.3 \text{ N/mm}^2$
Horizontal shearing force/key $=$ stress $\times$ horizontal area between keys
$$= 0.3 \times (700 \times 250) \times 10^{-3}$$
$$= 52.5 \text{ kN}$$

$$\text{Required sectional area of key} = \frac{\text{shearing force}}{\text{allowable average shear stress}}$$

$$= (52.5 \times 10^3)/2$$
$$= 26\,250 \text{ mm}^2$$

$\therefore$ Required thickness of key $=$ required area/width of key
$$= 26\,250/250$$
$$= \underline{105.0 \text{ mm}}$$

**Example 7.8**

The box section shown in Figure 7.16 is formed from two $305 \times 102$ channel sections connected by two $300 \text{ mm} \times 10 \text{ mm}$ cover plates. Each of the channel sections has a second moment of area about the neutral axis of $65.87 \times 10^6 \text{ mm}^4$. The cover plates are attached by 20 mm diameter bolts in pairs, for which the maximum permissible average shear stress is $95 \text{ N/mm}^2$.

The section is to be used as a cantilever, 4.0 m long and carrying a distributed load which varies linearly from $w$/unit length at the fixed end to zero at the free end. The pitch of the bolts is varied at stages along the cantilever, as shown in the figure. Based upon consideration of the shear strength of the bolts, what is the maximum value of $w$ that may be applied?

(Birmingham University)

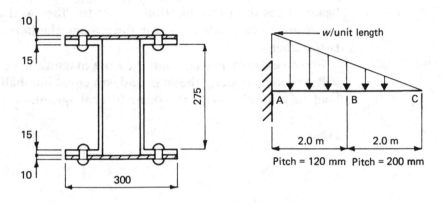

**Figure 7.16**

**This is a more complicated variation of the last problem, where the bending action of the compound beam depends on the ability of the bolts to resist the shear along the horizontal plane between the channels and the cover plates.**

*Solution 7.8*

First calculate the second moment of area of the compound section:

| Part | Area ($A$) (mm²) | $I_{CC}$ (mm⁴ × 10⁶) | $h = (z - \bar{z})$ (mm) | $Ah^2$ (mm⁴ × 10⁶) |
|------|------|------|------|------|
| One channel | – | 65.87 | – | 0 |
| One plate | 3000 | 0.03 | 157.5 | 74.42 |
| | | 65.90 | | 74.42 |

$$\therefore I_{XX} = 2 \times (65.90 + 74.42) \times 10^6 = \underline{280.64 \times 10^6 \text{ mm}^4}$$

The shear stress in the plate at the level of the interface between channel and plate is given by

$$\tau = \frac{QA\bar{y}}{Ib}$$

where the term $A\bar{y}$ is taken as the first moment of area of the plate about the neutral axis. Hence,

$$\tau = \frac{Q \times (3000 \times 157.5)}{(280.64 \times 10^6) \times 300} = 5.61 \times 10^{-6}\ Q\ \text{N/mm}^2$$

This is also the complementary shear stress along the horizontal face between channel and plate. Hence, the shearing *force* along 1 m of this horizontal face is given by

$$\begin{aligned}
\text{Shearing force} &= \text{shear stress} \times \text{area of shear surface} \\
&= 5.61 \times 10^{-6} \times Q \times (300 \times 1000) \times 10^{-3} \\
&= 1683.0 \times 10^{-6} \times Q\ \text{kN/metre length}
\end{aligned}$$

**Note that in the calculation of this shearing *force* we have first calculated the shear *stress* in the plate. We could have obtained the same answer by considering the stress in the channels at this level together with the plan area of the channels. Also note that $Q$ in the above equation for shearing force is expressed in newtons and in the subsequent calculations $Q$ is again expressed in newtons, to ensure the correct use of appropriate units.**

There are two critical sections, A and B, in the second diagram, where a critical shearing force is combined with a critical shear resistance due to the change in pitch of the bolts. The load per unit length that can be carried is $w$ kN/m and the shear strength of a bolt = cross-sectional area × allowable shear stress = $\pi \times (20^2/4) \times 95 \times 10^{-3} = 29.84$ kN. Taking the shearing force at a section as the area under the load distribution diagram to the right of the section, the calculations are as follows.

*At the section at B*

$$\text{Shearing force, } Q = \tfrac{1}{2} \times 2 \times \tfrac{1}{2}w = \tfrac{1}{2}w\ \text{kN}$$
$$\begin{aligned}
\text{Horizontal shearing force} &= 1683.0 \times 10^{-6} \times Q \\
&= 1683.0 \times 10^{-6} \times (\tfrac{1}{2}w \times 10^3) \\
&= 841.5 \times 10^{-3} \times w\ \text{kN}
\end{aligned}$$

$$\text{Shear resistance of bolts at 200 mm pitch} = 2 \times \left(29.84 \times \frac{1000}{200}\right) = 298.40\ \text{kN/m}$$

Equating shear resistance and horizontal shearing force,

$$841.5 \times 10^{-3} \times w = 298.40$$
$$\therefore w = 354.60\ \text{kN/m}$$

*At the section at A*

Shearing force, $Q = \frac{1}{2} \times 4 \times w = 2w$ kN

Horizontal shearing force $= 1683.0 \times 10^{-6} \times Q$

$\qquad = 1683.0 \times 10^{-6} \times (2w \times 10^3)$

$\qquad = 3366.0 \times 10^{-3} \times w$ kN

Shear resistance of bolts at 120 mm pitch $= 2 \times \left( 29.84 \times \dfrac{1000}{120} \right) = 497.33$ kN/m

Equating shear resistance and horizontal shearing force,

$$3366.0 \times 10^{-3} \times w = 497.33$$
$$\therefore w = 147.75 \text{ kN/m}$$

The *lowest* of these two answers gives the maximum load that can be carried. That is, $w = \underline{147.75 \text{ kN/m}}$.

## 7.5 Problems

**7.1** Figure P7.1 shows the cross-section of a solid beam which carries a vertical shearing force of 50 kN.

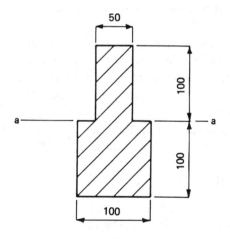

**Figure P7.1**

(i) Find the value of the shear stress just above and just below a–a and also at the centroid of the section. Compare the maximum stress in the section with the average shear stress.

(ii) Find the proportion of the shearing force which is carried by that part of the section which is above a–a.

(University of Portsmouth)

**7.2** Plot the shear stress distribution for the section shown in Figure P7.2 under a transverse shear force of 20 kN.

(Liverpool University)

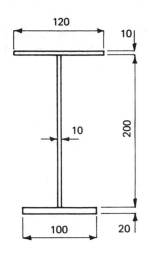

**Figure P7.2**

**7.3** Three equal channels are joined as shown in Figure P7.3(a) by 20 mm diameter bolts at 120 mm pitch. The allowable shear stress in the bolts is 80 N/mm². Each of the channels is 305 mm × 89 mm, with geometric properties as shown in Figure P7.3(b).

If the compound section is used as a beam with bending about the horizontal axis as shown, calculate the maximum shear that can be resisted, the bolts being the critical factor.

For each channel section take $A = 53.11$ cm², $I_{XX} = 7061$ cm⁴ and $I_{YY} = 5826$ cm⁴.

(University of Hertfordshire)

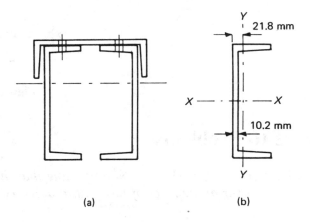

**Figure P7.3**

**7.4** A box-section beam is to be formed by welding together two 305 × 102 rolled channel sections as shown in Figure P7.4. Each channel section has cross-sectional area $A = 5883$ mm², second moment of area about a horizontal centroidal axis $I = 4.995 \times 10^6$ mm⁴, with the centroid 26.6 mm from the back of the web. If the assembled section is to carry a shear force of 350 kN in a

185

vertical direction, perpendicular to the webs of the channels, what will the shear force/unit length in each of the welds be?

(Birmingham University)

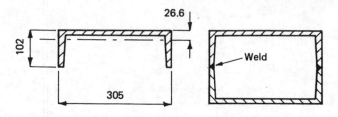

**Figure P7.4**

**7.5** The compound beam of Figure P7.5 is subjected to a uniformly distributed load of 500 kN/m over a span of 9 m. The channel is connected to the top flange of the I-section by two longitudinal welds at the toes of its flanges. Determine the longitudinal shear in each weld.

(Sheffield University)

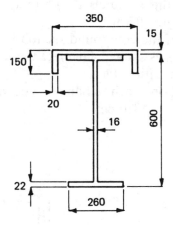

**Figure P7.5**

## 7.6 Answers to Problems

**7.1** (i) 7.27, 3.63, 3.78 N/mm² (*note that in this case the maximum stress does not occur at the level of the neutral axis*); (ii) 45.46%

**7.2**

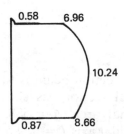

**Figure P7.6**

**7.3** 226.51 kN

**7.4** 1009.68 kN/m

**7.5** 1234.85 kN/m

# 8 Torsional Stress

## 8.1 Contents

The relationship between shear stress, torsional moment and angle of twist ● Thin-walled circular shafts ● Solid circular shafts ● Hollow circular shafts ● Circular shafts with change of section ● Non-prismatic shafts ● Power transmission.

This chapter deals with the development of shear stress due to the effect of applying torsional moments to solid or hollow shafts of circular cross-section. Examples are presented to illustrate the analysis or design of shafts under the action of torsional moments and also for the analysis or design of rotating shafts used to transmit power.

## 8.2 The Fact Sheet

### (a) Torsional Stress

Torsional stress results from the action of torsional or twisting moments acting about the longitudinal axis of a shaft. The effect of the application of a torsional moment, combined with appropriate fixity at the supports, is to cause torsional stresses within the shaft. These stresses, which are effectively shear stresses, are greatest on the outer surface of the shaft.

### (b) The Torsion Relationship

The relationship between the parameters that govern the torsion of a shaft is given by

$$\frac{\tau}{r} = \frac{T}{J} = \frac{G\theta}{L}$$

where $\tau$ = the shear stress at a distance $r$ from the central longitudinal axis of the shaft;

$r$ = the radius at which $\tau$ is being calculated;

$T$ = the torsional moment in the shaft;

$J$ = the polar second moment of area of the section about a longitudinal axis through the centroid of the section;

$G$ = the shear modulus (or modulus of rigidity) of the material of which the shaft is made;

$\theta$ = the angle of rotation; and

$L$ = the length of the shaft.

In applying the above equation to a thin-walled hollow circular shaft the stress $\tau$ can be assumed to be constant across the section and the *mean radius r* should be used in the formula.

## (c)   The Polar Second Moment of Area

For a thin-walled hollow circular shaft of mean radius $r$ and wall thickness $t$,

$$J = 2\pi r^3 t$$

For a solid circular shaft of radius $R$,

$$J = \frac{\pi R^4}{2}$$

For a hollow circular shaft of outer radius $R$ and inner radius $r$,

$$J = \frac{\pi}{2}(R^4 - r^4)$$

## (d)   Torsion of Non-prismatic Sections

The total angular rotation of a shaft of non uniform cross-section is given by:

$$\theta = \int_0^L \frac{T\,dx}{G\,J_x}$$

where

$J_x$ = the polar second moment of area at a distance $x$ from one end of the shaft; and

$L$ = the length of the shaft.

## (e)   Power–Torque Relationship

For a rotating shaft the power transmitted by the shaft is related to the torque developed within the shaft by the relationship

$$P = \frac{2\pi n T}{60} = \frac{\pi n T}{30}$$

where

$P$ = power (W);

$n$ = the number of revolutions per minute; and

$T$ = torque (Nm).

## 8.3 Symbols, Units and Sign Conventions

$G$ = the shear modulus of the material of which the shaft is made (kN/mm$^2$)

$J$ = the polar second moment of area of the section about a longitudinal axis through the centroid of the section (mm$^4$)

$L$ = the length of the shaft (mm)

$n$ = the number of revolutions per minute

$P$ = power (W)

$r$ = the radius at which $\tau$ is being calculated (mm)

$R$ = the outside radius of a shaft (mm)

$T$ = the torsional moment in the shaft (kNm)

$\tau$ = the shear stress at a distance $r$ from the central longitudinal axis of the shaft (N/mm$^2$)

$\theta$ = the angle of rotation (rad)

## 8.4 Worked Examples

### Example 8.1

A hollow circular shaft has inside and outside diameters of 30 mm and 40 mm, respectively. The material has a shear modulus of 60 kN/mm$^2$. Determine the maximum stress in the shaft when it is subjected to a twist of 5° in a length of 2.5 m. What is the maximum torque the shaft can carry?

(University of Westminster)

### Solution 8.1

**When tackling torsion problems, the torsion equation should first be written down and the appropriate part of the equation to be used should be determined by inspection:**

$$\frac{\tau}{r} = \frac{T}{J} = \frac{G\theta}{L}$$

**In the first part of the question the value of $\tau$ is to be determined. The outer diameter (and, hence, outer radius) is given. The twist (rotation) $\theta$ and length $L$ are also given. Therefore, by inspection the first and third group of terms in the above equation can be combined to give a solution for $\tau$.**

Hence,

$$\frac{\tau}{r} = \frac{G\theta}{L}$$

or

$$\tau = \frac{rG\theta}{L}$$

The maximum shear stress occurs at the outer surface of the shaft, where $r = 20$ mm. Hence,

$$\tau = \frac{20 \times (60 \times 10^3) \times (5 \times \pi/180)}{2500}$$

$$= 41.89 \text{ N/mm}^2$$

**Note that the units of the terms in the above equation are converted to newton and mm units. Most importantly, the rotation of 5° is converted to radians.**

To calculate the maximum torque the shaft can carry, use the second and third terms in the general expression. Hence

$$\frac{T}{J} = \frac{G\theta}{L}$$

or

$$T = \frac{GJ\theta}{L}$$

where

$$J = \frac{\pi}{2}(R^4 - r^4) = \frac{\pi}{2}(20^4 - 15^4) = 171.81 \times 10^3 \text{ mm}^4$$

$$\therefore T = \frac{(60 \times 10^3) \times (171.81 \times 10^3) \times (5 \times \pi/180)}{2500} \times \frac{1}{10^6}$$

$$= 0.36 \text{ kNm}$$

## Example 8.2

A round steel shaft ABC is shown in Figure 8.1. It is built-in at A and unsupported elsewhere. The part BC is 600 mm long and 35 mm in diameter. The part AB is 900 mm long and 50 mm in diameter, but a 30 mm diameter hole has been bored from the fixed end, A, to a depth of 300 mm. The shaft carries a pure torque, T, applied at the free end, C, and has been designed so that the

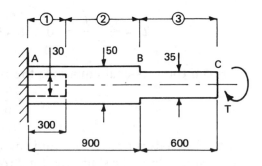

**Figure 8.1**

191

maximum shearing stress due to torsion is 90 N/mm². Determine the angle of twist at C resulting from the applied torque, $T$. The modulus of rigidity may be taken as 80 kN/mm².

(University of Portsmouth)

**The shaft can be considered to be constructed from three separate sections, marked 1, 2 and 3 in the figure. As the torque, $T$, is applied at the free end, the torsional moment will be constant along the shaft.**

### Solution 8.2

First calculate the polar second moment of area of each section of shaft:

Section 1    $J = \frac{1}{2}\pi(R^4 - r^4) = \frac{1}{2}\pi(25^4 - 15^4) = 534.07 \times 10^3 \text{ mm}^4$

Section 2    $J = \frac{1}{2}\pi R^4 = \frac{1}{2}\pi 25^4 = 613.59 \times 10^3 \text{ mm}^4$

Section 3    $J = \frac{1}{2}\pi R^4 = \frac{1}{2}\pi 17.5^4 = 147.32 \times 10^3 \text{ mm}^4$

**From the torsional equations the maximum torque that can be applied, for a given maximum shear stress, is given by**

$$T = \frac{\tau_{max}J}{R}$$

**where the maximum allowable shear stress, $\tau_{max}$, occurs on the outer surface of the shaft at a radius $R$ from the longitudinal axis. The torque is proportional to the ratio $J/R$ and, hence, the maximum torque that can be applied to this shaft will be dependent on the critical section of the shaft, where the ratio $J/R$ is _least_:**

Section 1    $\dfrac{J}{R} = \dfrac{534.07 \times 10^3}{25} = 21.36 \times 10^3 \text{ mm}^3$

Section 2    $\dfrac{J}{R} = \dfrac{613.59 \times 10^3}{25} = 24.54 \times 10^3 \text{ mm}^3$

Section 3    $\dfrac{J}{R} = \dfrac{147.32 \times 10^3}{17.5} = 8.42 \times 10^3 \text{ mm}^3$

Hence, the maximum torque that can be applied is governed by section 3 and is given by

$$T = \frac{\tau_{max}\,J}{R} = 90 \times 8.42 \times 10^3 \times 10^{-6} = \underline{0.76 \text{ kNm}}$$

From the torsional equations the rotation of a length of the shaft is given by

$$\theta = \frac{TL}{GJ}$$

As the shaft is fixed at one end and free at the other, and the torsional moment is constant along the shaft, the total rotation is the sum of the rotations of each of the three parts:

$$\theta = \sum \frac{TL}{GJ} = \frac{T}{G} \sum \frac{L}{J}$$

$$= \frac{0.76 \times 10^6}{80 \times 10^3} \left( \frac{300}{534.07 \times 10^3} + \frac{600}{613.59 \times 10^3} + \frac{600}{147.32 \times 10^3} \right)$$

$$= 0.053 \text{ rad}$$
$$= 180/\pi \times 0.053$$
$$= \underline{\underline{3.04°}}$$

## Example 8.3

A thin circular steel tube, thickness 2 mm and mean radius 50 mm, fits inside a thin circular brass tube, thickness 2 mm and mean radius 52 mm. The two tubes are connected together so that they act in combination.

If the maximum permissible shear stresses in the steel and brass are 80 N/mm$^2$ and 50 N/mm$^2$, respectively, find the maximum torque in kNm which may be applied to the combined tube. Take $G$ for steel $= 83 \times 10^3$ N/mm$^2$ and $G$ for brass $= 41 \times 10^3$ N/mm$^2$.

(Salford University)

**To solve this problem, it is necessary to determine which of the two limiting stresses governs the maximum torque that can be applied, because both limiting stresses are unlikely to be reached simultaneously. The two tubes are connected together, so that the rotation of each tube is the same. Hence, use can be made of an equation of** *displacement compatibility* **to find the relationship between the stresses in the two materials.**

## Solution 8.3

Using subscripts 'B' and 'S' to denote the brass and steel, respectively,

displacement compatibility,

$$\theta_B = \theta_S$$

But, from the torsional equations,

$$\theta_B = \left( \frac{\tau L}{GR} \right)_B$$

and

$$\theta_S = \left( \frac{\tau L}{GR} \right)_S$$

where for maximum shear stress in the steel $R_S = 50$ mm and for maximum shear stress in the brass $R_S = 52$ mm. Hence, combining the above equations,

$$\left(\frac{\tau L}{GR}\right)_B = \left(\frac{\tau L}{GR}\right)_S$$

$$\therefore \tau_B = \frac{G_B R_B}{G_S R_S} \times \tau_S$$

$$= \frac{41 \times 10^3 \times 52}{83 \times 10^3 \times 50} \times \tau_S$$

$$= 0.514 \times \tau_S$$

Assuming that the steel reaches its permissible shear stress of 80 N/mm² before the brass reaches its permissible shear stress, then

$$\tau_B = 0.514 \times 80 = 41.12 \text{ N/mm}^2$$

This is less than the permissible stress in the brass, of 50 N/mm², and, hence, the permissible steel stress governs the maximum torque that can be applied to this shaft.

Considering the *equilibrium condition* for the shaft, the total torque that can be carried is the sum of the torsional moments that can be carried by each of the two shaft components:

$$T = T_S + T_B$$

$$= (\tau \times J/R)_S + (\tau \times J/R)_B$$

$$= \left(\frac{\tau \times 2\pi R^3 t}{R}\right)_S + \left(\frac{\tau \times 2\pi R^3 t}{R}\right)_B$$

$$= \left\{\left(\frac{80.0 \times 2\pi \times 50^3 \times 2}{50}\right) + \left(\frac{41.12 \times 2\pi \times 52^3 \times 2}{52}\right)\right\} \times 10^{-6}$$

$$= \underline{3.91 \text{ kNm}}$$

**Note that this is a problem involving the use of thin-walled tubes. Hence, in the *equilibrium equation* the polar second moment of area of each part of the shaft has been calculated by use of the thin-tube formula given in the Fact Sheet.**

### Example 8.4

A steel bar, 1.0 m long, has a circular cross-section 60 mm diameter and is clamped against rotation at both ends. A concentrated torque of 12 kNm is applied at a distance 0.35 m from one end as shown in Figure 8.2. Assuming linear elastic behaviour, what are the fixing couples at the ends of the bar and what is the maximum shear stress in the steel?

(Birmingham University)

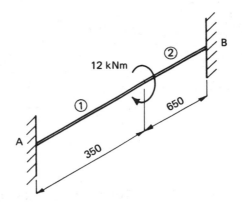

**Figure 8.2**

### Solution 8.4

Subscripts 1 and 2 are used to refer to the two lengths of the bar.

Because the two lengths of bar are clamped at their remote ends, the rotation of each length under the action of the applied torque will be the same. The *compatibility condition* for the bar is, therefore, given by:

$$\theta_1 = \theta_2 \tag{1}$$

But, from the torsional equations,

$$\theta_1 = \frac{T_1 L_1}{GJ} \quad \text{and} \quad \theta_2 = \frac{T_2 L_2}{GJ} \tag{2}$$

where $T_1$, and $T_2$ are the torsional moments in each length of the bar.

**If the equilibrium of each length of the bar is considered, it will be appreciated that $T_1$ and $T_2$ are also the respective fixing couples at the clamped ends.**

Hence, combining Equations (1) and (2),

$$\frac{T_1 L_1}{GJ} = \frac{T_2 L_2}{GJ}$$

or

$$T_1 L_1 = T_2 L_2 \tag{3}$$

The *equilibrium condition* for the bar states that the sum of the internal torsional moments must equal the externally applied torque. That is,

$$T_1 + T_2 = 12 \text{ kNm} \tag{4}$$

Hence, combining Equation (3) with Equation (4) to eliminate $T_2$,

$$T_1 + \frac{T_1 L_1}{L_2} = 12$$

$$\therefore \frac{T_1 L_2 + T_1 L_1}{L_2} = 12$$

or

$$T_1 = 12 \frac{L_2}{(L_1 + L_2)}$$

$$= 12 \times \frac{650}{(350 + 650)}$$

$$= 7.8 \text{ kNm}$$

i.e.

fixing moment at end A of the bar = 7.8 kNm

And, from Equation (4),

$$T_2 = 12 - T_1$$
$$= 12 - 7.8$$
$$= 4.2 \text{ kNm}$$

i.e.

fixing moment at end B of the bar = 4.2 kNm

The maximum shear stress occurs along the length of bar where the torsional moment is largest and at the maximum radius. Hence, from the torsional equations,

$$\tau = \frac{RT}{J}$$

where, for a solid circular section,

$$J = \tfrac{1}{2}\pi R^4 = \tfrac{1}{2}\pi 30^4 = 1.272 \times 10^6 \text{ mm}^4$$

Therefore,

$$\tau = \frac{30 \times 7.8 \times 10^6}{1.272 \times 10^6}$$

$$= 183.96 \text{ N/mm}^2$$

## Example 8.5

A steel pipe is to be constructed in three lengths, AB, BC and CD, as shown in Figure 8.3. Each length has a thin-walled circular cross-section 350 mm mean diameter; AB and CD have a wall thickness 12 mm; BC has a wall thickness 10 mm. The two end lengths are rigidly fixed against rotation at A and D; the centre length, BC, is to be inserted and attached by flanges welded to the ends and drilled to accept the connecting bolts.

During assembly it is found one flange on BC has been misaligned by 1.5°, so that a torque must be applied before bolts can be inserted.

(a) What is the torque required and what is the shear stress in BC when the torque is applied?

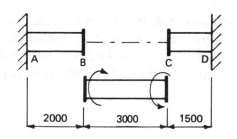

**Figure 8.3**

(b)   What are the shear stresses in each length when the bolts are inserted and the torque removed? (Take $G = 80$ kN/mm².)

(Birmingham University)

### Solution 8.5

(a)
**All sections of this pipe are described in the question as thin sections and therefore the equation for the polar second moment of area of a thin-walled circular section can be used in answering this problem.**

For length BC,

$$J = 2\pi r^3 t = 2 \times \pi \times 175^3 \times 10 = 336.74 \times 10^6 \text{ mm}^4$$

For lengths AB and CD,

$$J = 2\pi r^3 t = 2 \times \pi \times 175^3 \times 12 = 404.09 \times 10^6 \text{ mm}^4$$

$$\text{Angle of rotation of BC, } \theta = 1.5° = \frac{1.5 \times \pi}{180} = 0.0262 \text{ rad}$$

From the torsional equation applied to BC,

$$T = \frac{JG\theta}{L} = \frac{336.74 \times 10^6 \times 80 \times 10^3 \times 0.0262}{3000}$$

$$= 235.27 \times 10^6 \text{ Nm}$$
$$= \underline{235.27 \text{ kNm}}$$

The corresponding shear stress in BC is given by

$$\tau = \frac{G\theta r}{L} = \frac{80 \times 10^3 \times 0.0262 \times 175}{3000}$$

$$= \underline{122.27 \text{ N/mm}^2}$$

(b)
**This question can be tackled using the *principle of superposition*. Figure 8.4 shows the pipe under the action of a set of external torsional moments applied at B and C, equal and opposite to those calculated in part (a) of the question and with the lengths of the pipe assumed to be fastened together. If the system of forces,**

**stresses and displacements resulting from the loading indicated in Figure 8.4 is *superimposed* on to that indicated in Figure 8.3, then the external torsional moments will cancel out and the final shear stresses in each length of the pipe will be given by the addition of the respective stresses from both part (a) and part (b).**

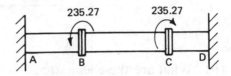

**Figure 8.4**

Considering the system shown in Figure 8.4, the *compatibility* equations for the pipe can be obtained by considering the rotation of joint B ($\theta_B$) and joint C ($\theta_C$):

for length AB, $\theta_{AB} = \theta_B$
for length BC, $\theta_{BC} = \theta_B + \theta_C$
for length CD, $\theta_{CD} = \theta_C$

Hence, from the torsional equations,

$$T_{AB} = \frac{GJ_{AB}}{L_{AB}} \theta_B \qquad T_{BC} = \frac{GJ_{BC}}{L_{BC}} (\theta_B + \theta_C) \qquad T_{CD} = \frac{GJ_{CD}}{L_{CD}} \theta_C$$

Rearranging these equations,

$$\theta_B = \frac{T_{AB}L_{AB}}{GJ_{AB}} \qquad (\theta_B + \theta_C) = \frac{T_{BC}L_{BC}}{GJ_{BC}} \qquad \theta_C = \frac{T_{CD}L_{CD}}{GJ_{CD}}$$

from which,

$$(\theta_B + \theta_C) = \frac{T_{BC}L_{BC}}{GJ_{BC}} = \frac{T_{AB}L_{AB}}{GJ_{AB}} + \frac{T_{CD}L_{CD}}{GJ_{CD}} \tag{1}$$

The *equilibrium equations* for the pipe can be obtained by considering the equilibrium of both joint B and joint C, where the externally applied torque ($T = 235.27$ kNm) must equal the sum of the internal torsional moments.

At joint B, $T = T_{AB} + T_{BC}$
At joint C, $T = T_{BC} + T_{CD}$

Hence,

$$T_{AB} + T_{BC} = T_{BC} + T_{CD}$$

or

$$T_{AB} = T_{CD} \tag{2}$$

and

$$T_{BC} = T - T_{AB} \tag{3}$$

198

Hence, substituting Equations (2) and (3) into Equation (1),

$$\frac{(T - T_{AB})\, L_{BC}}{GJ_{BC}} = \frac{T_{AB}L_{AB}}{GJ_{AB}} + \frac{T_{AB}L_{CD}}{GJ_{CD}}$$

which rearranges to give

$$T_{AB}\left(\frac{L_{AB}}{J_{AB}} + \frac{L_{BC}}{J_{BC}} + \frac{L_{CD}}{J_{CD}}\right) = T\left(\frac{L_{BC}}{J_{BC}}\right)$$

$$\therefore \frac{T_{AB}}{2\pi r^3}\left(\frac{L_{AB}}{t_{AB}} + \frac{L_{BC}}{t_{BC}} + \frac{L_{CD}}{t_{CD}}\right) = \frac{T}{2\pi r^3}\left(\frac{L_{BC}}{t_{BC}}\right)$$

$$\therefore T_{AB}\left(\frac{2000}{12} + \frac{3000}{10} + \frac{1500}{12}\right) = 235.27\left(\frac{3000}{10}\right)$$

from which,

$$T_{AB} = \underline{119.29 \text{ kNm}}\ (= T_{CD})$$

and, from Equation (3),

$$\begin{aligned} T_{BC} &= T - T_{AB} \\ &= 235.27 - 119.29 \\ &= \underline{115.98 \text{ kNm}} \end{aligned}$$

Hence, calculating the shear stress in each section of the pipe and *algebraically* adding the calculated stresses to the corresponding stresses from part (a) of this problem,

$$\text{In length AB, } \tau = \frac{rT}{J} = \frac{175 \times 119.29 \times 10^6}{404.09 \times 10^6} - 0 = 51.66 \text{ N/mm}^2$$

$$\text{In length BC, } \tau = \frac{rT}{J} = \frac{175 \times 115.98 \times 10^6}{336.74 \times 10^6} - 122.27 = -62.00 \text{ N/mm}^2$$

$$\text{In length CD, } \tau = \frac{rT}{J} = \frac{175 \times 119.29 \times 10^6}{404.09 \times 10^6} - 0 = 51.66 \text{ N/mm}^2$$

## Example 8.6

A solid circular steel shaft 2 m long tapers uniformly from 100 mm diameter to 60 mm diameter. One end is held rigid while the other is subjected to a torque of 4 kNm. Find the twist (in degrees) and the maximum shear stress in the shaft. Take the shear modulus of steel as $83 \times 10^3$ N/mm$^2$.

(Salford University)

**The shaft has been drawn in Figure 8.5, which shows a section $X$–$X$ at distance $x$ from one end of the shaft. It is necessary to calculate the radius at this section, and, hence, the polar second moment of area at this section, which can then be used in the general expression for the torsion of a tapered shaft. (See the Fact Sheet.)**

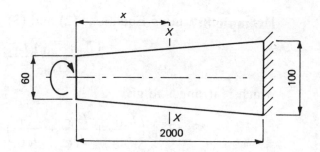

**Figure 8.5**

*Solution 8.6*

At section $X-X$,

$$R = \left\{ 30 + \frac{(20 \times x)}{2000} \right\} = \frac{(3000 + x)}{100} \text{ mm}$$

Therefore,

polar second moment of area, $J_x = \frac{1}{2}\pi R^4 = \frac{1}{2}\pi \dfrac{(3000 + x)^4}{100^4}$

Rotation of shaft, $\theta = \displaystyle\int_0^L \frac{T\,dx}{GJ_x}$

$$= \int_0^{2000} \frac{(4 \times 10^6) \times 100^4}{83 \times 10^3 \times \frac{1}{2}\pi(3000 + x)^4} \, dx$$

$$= 3.068 \times 10^9 \int_0^{2000} \frac{1}{(3000 + x)^4} \, dx$$

$$= 3.068 \times 10^9 \left[ -\frac{1}{3} \frac{1}{(3000 + x)^3} \right]_0^{2000}$$

$$= -\frac{3.068 \times 10^9}{3} \left[ \frac{1}{5000^3} - \frac{1}{3000^3} \right]$$

$$= 0.0297 \text{ rad}$$
$$= 0.0297 \times 180/\pi$$
$$= \underline{1.70°}$$

The largest shear stress occurs at the smaller of the two ends and, using the torsional equation,

$$\tau = \frac{TR}{J}$$

$$= \frac{4 \times 10^6 \times 30}{\frac{1}{2}\pi \times 30^4}$$

$$= \underline{94.31 \text{ N/mm}^2}$$

## Example 8.7

Two circular shafts transmitting a torque $T$ are connected by a flanged coupling having 6 bolts of 20 mm diameter, as shown in Figure 8.6. If the average shear stress on a bolt must not exceed 70 N/mm² and the maximum shear stress in the shaft must not exceed 120 N/mm², determine the maximum value for the torque $T$ and the maximum power that could be transmitted at 390 revolutions per minute.

(Nottingham University)

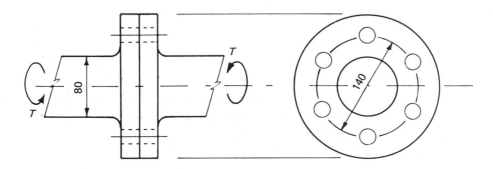

**Figure 8.6**

**There are two limiting conditions specified for this problem: a limiting shear stress in the bolts and a limiting shear stress in the shaft itself. Hence, it is necessary to examine both limiting stress conditions and decide which one governs the maximum torque that can be applied.**

*Solution 8.7*

Considering shear in the shaft and using the torsional equation:

$$T = \frac{\tau J}{R}$$

where

$$J = \tfrac{1}{2}\pi R^4 = \tfrac{1}{2}\pi \times 40^4 = 4.02 \times 10^6 \text{ mm}^4$$

$$\therefore T = \frac{120 \times 4.02 \times 10^6}{40} \times 10^{-6}$$

$$= 12.06 \text{ kNm}$$

Considering shear in the bolts:

The shearing force ($F$) in one bolt when the limiting shear stress is reached is given by

201

$$F = \text{limiting shear stress} \times \text{area of bolt section}$$
$$= \tau \times (\pi \times d^2/4)$$
$$= 70 \times \pi \times 20^2/4$$
$$= 21991.13 \text{ N}$$
$$= 21.99 \text{ kN}$$

The torsional capacity of the bolted connection can be determined by taking moments of the limiting shearing force ($F$) in each bolt about the central longitudinal axis of the shaft. The sum of the moments of the forces in all six bolts will give the total torsional capacity of the bolted connection.

Hence,

$$T = 6 \times (F \times r)$$
$$= 6 \times 21.99 \times 0.070$$
$$= 9.24 \text{ kNm}$$

As this is the lower value obtained from the two calculations, the bolted connection governs the strength of the shaft and the maximum torque that can be applied is 9.24 kNm.

The power ($P$) that can be transmitted can be calculated from the standard equation given in the Fact Sheet:

$$P = \frac{T \pi n}{30}$$

$$= \frac{9.24 \times 10^3 \times \pi \times 390}{30}$$

$$= 377\,367.79 \text{ W}$$
$$= 377.37 \text{ kW}$$

## Example 8.8

A hollow circular shaft is required to transmit power of 400 kW at 420 revolutions per minute. The outside and inside diameters of the 600 mm long shaft are chosen such that:

(i) the maximum shear stress in the shaft is 90 N/mm$^2$;
(ii) the flexibility of the shaft (expressed as twist per unit torque) is 0.070 degrees per kNm.

(a) Determine the torque to be transmitted by the shaft.
(b) Show that the polar second moment of area of the shaft must be approximately 6.1 × 10$^6$ mm$^4$.
(c) Determine the necessary external and internal diameters of the shaft to satisfy the above conditions.

($G = 80$ kN/mm$^2$.)

(Nottingham University)

## Solution 8.8

The torque to be transmitted can be calculated from the standard torque–power relationship which is to be found in the Fact Sheet:

$$P = \frac{T\pi n}{30}$$

or

$$T = \frac{30P}{\pi n}$$

$$= \frac{30 \times (400 \times 10^3)}{\pi \times 420} \times 10^{-3}$$

$$= 9.09 \text{ kNm}$$

(b)

Shaft flexibility = 0.07 deg/kNm

Hence,

$$\text{rotation at maximum torque} = 0.07 \times 9.09$$
$$= 0.636°$$
$$= 0.636 \times \pi/180$$
$$= 0.0111 \text{ rad}$$

The maximum torque and maximum rotation are related by the torsional equation. That is,

$$\frac{T}{J} = \frac{G\theta}{L}$$

Hence, the required polar second moment of area, $J$, is given by

$$J = \frac{TL}{G\theta}$$

$$= \frac{9.09 \times 10^6 \times 600}{80 \times 10^3 \times 0.0111}$$

$$= 6.14 \times 10^6 \text{ mm}^4$$

(c)  Let the external and internal radii of the shaft be $R$ and $r$, respectively: Hence, from the calculation in part (b),

$$J = \tfrac{1}{2}\pi(R^4 - r^4) = 6.14 \times 10^6 \text{ mm}^4 \qquad (1)$$

This equation is insufficient to determine the geometry of the shaft, and one additional equation must be found to enable $R$ and $r$ to be calculated. So far no use has been made of the information specifying the maximum shear stress in the shaft. If use is made of this limiting condition, an additional equation can be written down.

The maximum shear stress occurs on the outer surface of the shaft at maximum radius. Hence, from the torsional equations,

$$\frac{\tau}{R} = \frac{T}{J}$$

or

$$R = \frac{\tau J}{T}$$

$$= \frac{90 \times 6.14 \times 10^6}{9.09 \times 10^6}$$

$$= 60.792 \text{ mm}$$

Hence, from Equation (1),

$$J = \tfrac{1}{2}\pi(60.792^4 - r^4) = 6.14 \times 10^6 \text{ mm}^4$$

which solves to give $r = 55.878$ mm. Hence,

$$\text{external diameter} = 2 \times R = 2 \times 60.792 = \underline{121.58 \text{ mm}}$$

$$\text{internal diameter} = 2 \times r = 2 \times 55.878 = \underline{111.76 \text{ mm}}$$

## 8.5 Problems

**8.1** A shaft AB of length 5 m has a hollow circular cross-section of 100 mm outside diameter and 50 mm inside diameter. An axial torque of 20 kNm is applied to the shaft at point C, 2 m from end A.

Calculate the maximum shear stress developed in the material of the shaft and the angle of rotation of the shaft at point C for each of the following restraint conditions:

(a) end B restrained against rotation and end A free;
(b) ends A and B both restrained against rotation.

Modulus of rigidity $G = 80 \text{ kN/mm}^2$.

(Nottingham Trent University)

**8.2** A drive shaft designed for operation in an aggressive environment consists of two concentric thin-walled tubes which are fixed to each other by stiff discs at their ends, as shown in Figure P8.1. The outer tube, which is intended to provide corrosion protection, is of aluminium and has a mean diameter of 100 mm and a wall thickness of 3 mm. The inner tube is of steel.

Find the mean diameter and wall thickness of the inner tube such that, when an axial torque is applied to the assembly, the consequent shear stresses in the steel and the aluminium are in the ratio 2 to 1 and the torque is shared equally between the two tubes.

For steel $G_S = 83 \text{ kN/mm}^2$ and for aluminium $G_A = 28 \text{ kN/mm}^2$.

(Manchester University)

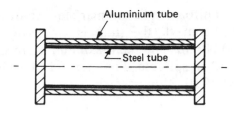

**Figure P8.1**

8.3 A steel bar ABCD with a solid circular cross-section 75 mm diameter is fixed at both ends and carries concentrated torques of 3 kNm and 7 kNm at B and C, respectively, as shown in Figure P8.2. What are the reactions at the end of the bar, what is the maximum shear stress in the steel and what are the rotations at B and C? ($G = 80$ kN/mm$^2$.)

(Birmingham University)

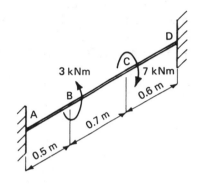

**Figure P8.2**

8.4 The circular shaft ABC shown in Figure P8.3 is rigidly built-in at A and C. The portion AB is a solid shaft of 20 mm diameter. The portion BC is a hollow shaft of 20 mm outer diameter and 10 mm inner diameter. The material of both portions is steel, which has a shear modulus of 83 kN/mm$^2$. If a torque of 200 Nm is applied at B, find:

(a) how much of this torque is carried by each of AB and BC;
(b) the maximum shear stress at any point;
(c) the amount of twist in degrees at B.

(University of Hertfordshire)

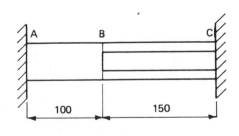

**Figure P8.3**

**8.5** A uniform solid circular steel shaft is required to transmit 300 kW of power at 500 rpm. If the shaft is 2 m long, the maximum permissible shear stress is 60 N/mm$^2$ and the maximum permissible twist over the length of the shaft is 1°, find a suitable diameter for the shaft. Take the shear modulus of steel as $83 \times 10^3$ N/mm$^2$.

(Salford University)

## 8.6 Answers to Problems

**8.1** (a) 108.65 N/mm$^2$, 4.67°; (b) 65.19 N/mm$^2$, 1.87°

**8.2** 67.47 mm, 3.30 mm

**8.3** Fixing moment at A = 3.84 kNm, fixing moment at D = 0.16 kNm (both moments anticlockwise when viewed in the direction from A to D); 46.36 N/mm$^2$; rotation at B = 0.02°, rotation at C = 0.53° (both rotations clockwise when viewed in the direction from A to D)

**8.4** (a) 123.08 Nm, 76.92 Nm; (b) 78.36 N/mm$^2$; (c) 0.54°

**8.5** 94.74 mm

# 9 Mohr's Circles of Stress and Strain

## 9.1 Contents

Mohr's circle of stress — construction and interpretation ● Principal stresses ● Principal planes ● Maximum shear stress ● Mohr's circle of strain — construction and interpretation ● Principal strains ● Strain gauge rosettes.

Most of the questions in this chapter can be answered either by graphically constructing a Mohr's circle or by using the equations that describe the circle and the critical points on the circle. The Mohr's circle gives a complete visual picture of the state of stress (or strain) at a point in a structure and is often easier to use and interpret than a set of equations. For this reason graphical solutions are given to all the questions, although the reader may wish to check the validity of the answers by producing solutions using the equations.

As the required answers in most cases necessitate only a graphical construction with appropriate interpretation of the Mohr's circle, a large part of the given solutions consists of explanation describing the circle construction. In an examination such written explanation would not be necessary, but the student should ensure that the geometrical construction used to draw the circle is clearly shown and that the required answers are marked clearly on the circle and summarised in writing separately.

## 9.2 The Fact Sheet

### (a) Mohr's Circle of Stress

Given an element of material representing a point in a structure, as shown in Figure 9.1, with known normal stresses ($\sigma_x$ and $\sigma_y$) and shear stress ($\tau'$), the Mohr's circle of stress, as shown in Figure 9.2, can be drawn as follows:

(i) To some suitable scale plot the point A with coordinates ($\sigma_x$, $\tau'$) representing the state of stress on the faces of the element which are parallel to the reference $Y$ axis.

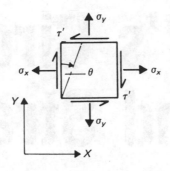

**Figure 9.1**

(ii) Plot the point B with coordinates $(\sigma_y, -\tau')$ representing the state of stress on the faces of the element which are parallel to the reference $X$ axis.

(iii) Join A to B and locate the centre of the circle where the line AB intersects the $\sigma$ axis at H. This point has coordinates $(\{\sigma_x + \sigma_y\}/2, 0)$

(iv) With a compass centred at H draw a circle passing through A and B. This is the Mohr's circle of stress, which describes the state of stress at the point under consideration.

## (b)  Interpretation of the Mohr's Circle of Stress

To determine the state of stress on a plane inclined at an angle $\theta$ as shown in Figure 9.1, where $\theta$ is measured in a *clockwise* direction from the $Y$ axis:

(i) From the line HA in Figure 9.2 turn off an angle $2\theta$ in a clockwise direction and draw the line HG.

(ii) Read the coordinates $(\sigma, \tau)$ of G to give the normal stress, $\sigma$, and shear stress, $\tau$, on the plane inclined at $\theta$ to the $Y$ axis.

Any two points at the opposite end of a diagonal of the circle represent the state of stress on two mutually perpendicular planes.

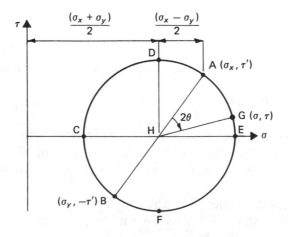

**Figure 9.2**  Mohr's circle of stress

### (c) Principal Stresses and Principal Planes

The *principal stresses* are the maximum and minimum normal stresses that occur at a point within a structure, and they act on the *principal planes*, which are orientated at 90° to each other. The maximum and minimum principal stresses are represented by the points E and C, respectively, in Figure 9.2. No shear stresses act on the principal planes.

### (d) Maximum Shear Stresses

The points D and F represent the state of stress on two mutually perpendicular planes where the shear stresses are a maximum. The planes of maximum shear stress are orientated at 45° to the principal planes.

### (e) The Pole Point Construction

The pole point construction can be used to determine the inclination of planes on which known stresses act or to determine the stresses acting on planes of known inclination. Using a previously constructed Mohr's circle of stress as shown in Figure 9.3, the following geometric construction can be used:

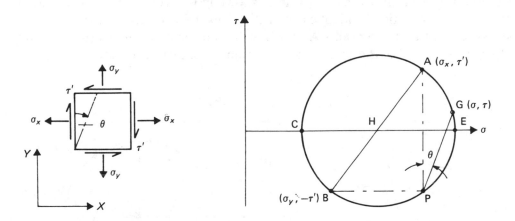

**Figure 9.3**

(i) From A, which represents the stresses $\sigma_x$ and $\tau'$ on the faces of the element which are parallel to the reference $Y$ axis, draw a line *parallel to the Y axis* to meet the circle at P. P is known as the *pole point*. Alternatively, P could be located by drawing a line from B parallel to the reference $X$ axis, as B represents the state of stress on the faces of the element which are parallel to the $X$ axis.

(ii) To determine the state of stress on a plane of known inclination $\theta$ (measured clockwise from the $Y$ axis), through P draw a line parallel to the specified plane (i.e. inclined at the same angle, $\theta$). Where this line intersects the circle at G, the coordinates of G will give the stresses, $\sigma$ and $\tau$, on the plane.

(iii) To determine the inclination of a plane on which known stresses act, determine the point on the circle representing these known stresses and connect this point to P. The slope of the line will give the inclination of the plane. For example, connecting E to P and C to P will give the orientation of the principal planes.

## (f) Formulae for Principal Stresses and Maximum Shear Stresses

For the element of material shown in Figure 9.1:

$$\text{maximum shear stress } (\tau_{max}) = \pm [\{0.5(\sigma_x - \sigma_y)\}^2 + \tau'^2]^{1/2}$$
$$\text{maximum principal stress } (\sigma_{max}) = 0.5(\sigma_x + \sigma_y) + [\{0.5(\sigma_x - \sigma_y)\}^2 + \tau'^2]^{1/2}$$
$$\text{minimum principal stress } (\sigma_{min}) = 0.5(\sigma_x + \sigma_y) - [\{0.5(\sigma_x - \sigma_y)\}^2 + \tau'^2]^{1/2}$$

plane of maximum principal stress is given by

$$\tan(2\theta) = \frac{2\tau'}{(\sigma_x - \sigma_y)}$$

## (g) Mohr's Circle of Strain

The state of strain indicated on the small element shown in Figure 9.4 can be represented by a Mohr's circle of strain as shown in Figure 9.5. The construction of the circle is practically identical with the construction of the Mohr's circle of stress, except that the points A and B are given the coordinates $(\epsilon_x, \frac{1}{2}\gamma')$ and $(\epsilon_y, -\frac{1}{2}\gamma')$, where $\gamma'$ is the angular shear strain, as shown in Figure 9.4.

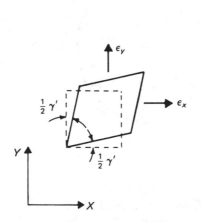

**Figure 9.4**

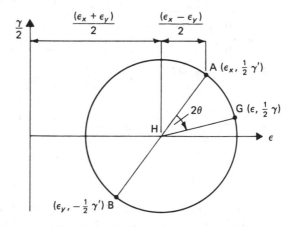

**Figure 9.5** Mohr's circle of strain

The interpretation of the circle is identical with the interpretation of the Mohr's circle of stress, except that it should be remembered that the vertical coordinates are half the value of shear strain, whereas in the circles of stress the vertical coordinates represent the full value of shear stress.

# 9.3   Symbols, Units and Sign Conventions

$E =$ Young's modulus of elasticity (kN/mm$^2$)

$\epsilon =$ strain in a direction normal to a plane

$\epsilon_x =$ strain in a direction normal to a plane which is parallel to the reference $Y$ axis

$\epsilon_y =$ strain in a direction normal to a plane which is parallel to the reference $X$ axis

$\sigma =$ normal stress acting on a plane (N/mm$^2$)

$\sigma_x =$ stress normal to a plane which is parallel to the reference $Y$ axis (N/mm$^2$)

$\sigma_y =$ stress normal to a plane which is parallel to the reference $X$ axis (N/mm$^2$)

$\tau =$ shear stress acting on a plane (N/mm$^2$)

$\tau' =$ shear stress and complementary shear stress acting on the faces of a rectangular element (N/mm$^2$)

$\nu =$ Poisson's ratio

$\theta =$ angle between a plane and the positive direction of the reference $Y$ axis, measured in a clockwise direction

$\gamma =$ angular shear strain

Tensile strains are taken as positive.
Tensile stresses are taken as positive.
Shear stresses acting on opposite faces of an element and which form a clockwise couple are taken as positive.
The angle $\theta$ is measured in a clockwise direction.

# 9.4   Worked Examples

### Example 9.1

An isolated element of material cut from a certain structure has direct and shear stress components as shown in Figure 9.6. It is required to find direct and shear stress components $\sigma$ and $\tau$ acting on a plane inclined at an angle $\theta$ as defined in the diagram.

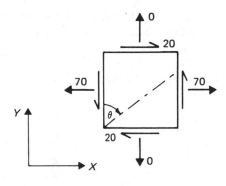

**Figure 9.6**

(a)   Draw a suitable triangular element, and by considering the equilibrium of forces acting on it, obtain expressions for $\sigma$ and $\tau$ in terms of $\theta$. On a Mohr's

211

plot of $\tau$ against $\sigma$ mark points corresponding to $\theta = 0$, $45°$, $90°$ and $135°$ and hence sketch the entire circular $\sigma$, $\tau$ locus.

(b)  Hence determine:

    (i)   values of $\theta$ for planes of maximum $\tau$ and corresponding values of $\tau$;
    (ii)  values of $\theta$ for planes of maximum and minimum $\sigma$ and the corresponding values of $\sigma$ (i.e. principal planes and stresses);
    (iii) values of $\theta$ for planes of zero $\sigma$.

(Cambridge University)

### Solution 9.1

(a)  Consider the stresses acting on the triangular element shown in Figure 9.7. If the area of face AB is $A$, then the *forces* acting on the face of the element are as shown in Figure 9.8. Hence, resolving forces perpendicular to AB,

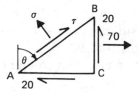

**Figure 9.7**  Stresses

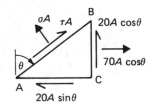

**Figure 9.8**  Forces

$$\sigma \times A = (70 \times A \cos \theta) \times \cos \theta - (20 \times A \cos \theta) \times \sin \theta$$
$$- (20 \times A \sin \theta) \times \cos \theta$$
$$\therefore \sigma = 70 \times \cos^2 \theta - 40 \times \sin \theta \times \cos \theta$$
$$= 70 \times \cos^2 \theta - 20 \times \sin (2\theta) \qquad (1)$$

And resolving forces parallel to AB:

$$\tau \times A = (20 \times A \sin \theta) \times \sin \theta - (20 \times A \cos \theta) \times \cos \theta$$
$$- (70 \times A \cos \theta) \times \sin \theta$$
$$\therefore \tau = 20 (\sin^2 \theta - \cos^2 \theta) - 70 (\sin \theta \times \cos \theta)$$
$$= -20 \times \cos (2\theta) - 35 \times \sin (2\theta) \qquad (2)$$

Hence, using Equations (1) and (2) and the given values of $\theta$:

| $\theta$ | $\sigma$ | $\tau$ | |
|---|---|---|---|
| 0 | 70 | −20 | (A) |
| 45 | 15 | −35 | (K) |
| 90 | 0 | 20 | (B) |
| 135 | 55 | 35 | (L) |

    **These four points are plotted to scale as shown in Figure 9.9. The two points marked A and B represent the state of stress on two mutually perpendicular planes. The two points therefore lie on opposite ends of a diagonal. To draw the**

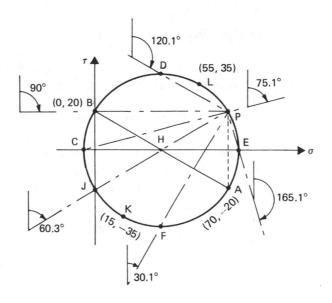

**Figure 9.9**

circle, join A to B, locate the centre of the circle where this line crosses the $\sigma$ axis at H and with a compass centred at H and with radius AH draw the circle. Points K and L could similarly be used to locate H.

(b)
The point marked A represents the state of stress on a plane which is parallel to the reference $Y$ axis ($\theta = 0°$). The pole point can be located by drawing a line through A, parallel to the plane ($\theta = 0°$) to meet the circle at P. This locates the Pole point.

(i)   The maximum shear stresses are given by the points marked D and F. To determine the inclination of the planes of maximum shear stress, join P to both D and F. The inclination of these lines gives the inclination of the planes of maximum shear stress, which, together with the values of maximum shear stress, can be scaled off the diagram.

$$\text{Maximum shear stress} = \pm 40.3 \text{ N/mm}^2: \quad \theta = 120.1°; \; 30.1°$$

(ii)   The principal stresses are given by the points marked E and C. Again join P to both E and C, to obtain the inclination of the principal planes, and scale the inclinations and the value of the principal stresses from the diagram.

$$\text{Maximum principal stress} = +75.3 \text{ N/mm}^2: \quad \theta = 165.1°$$

$$\text{Minimum principal stress} = -5.3 \text{ N/mm}^2: \quad \theta = \phantom{1}75.1°$$

(iii)   The planes of zero normal stress are indicated by the points marked B and J on the diagram. Join P to both B and J and measure the inclination of these lines as shown.

$$\text{For planes of zero } \sigma: \theta = 90°; \quad \theta = 60.3°$$

213

## Example 9.2

The direct and shear stresses shown in Figure 9.10 are developed at a point in a steel structure. At this point determine:

(i) the inclination of the principal planes and the principal stresses which act on them;
(ii) the normal and shear stresses acting on the plane AB;
(iii) the direct strain acting along the direction AB if the modulus of elasticity, $E = 210$ kN/mm$^2$ and Poisson's ratio $v = 0.28$.

The answers to part (i) and (ii) should be shown clearly on a diagram of an element.

(Nottingham Trent University)

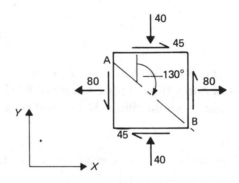

**Figure 9.10**

## Solution 9.2

(i), (ii)
**To draw the Mohr's circle, plot point A (+80, −45) and point B (−40, +45), representing the stress conditions on the faces parallel and perpendicular to the reference $Y$ axis. Note that the signs of the stresses are consistent with the convention indicated in Section 9.3. Join A to B and where this line intersects the $\sigma$ axis locate the centre of the circle at H. With compass centred at H and with radius AH draw the circle.**

**To locate the pole point, through A draw a line parallel to the reference $Y$ axis, as A represents the stress condition on a plane parallel to the $Y$ axis. P is the point where this line intersects the circle.**

**To answer part (i) of the question, join P to both E and C, which represent the stress conditions on the principal planes. Scale the value of the principal stresses from the diagram and measure the inclination of the lines PE and PC. The answers are shown on the diagram on a correctly orientated element which is inclined such that the faces are parallel to the principal axes.**

**To answer part (ii) of the question, through P draw a line parallel to the plane AB in the original figure and, hence, inclined at 130°. The point where this line intersects the circle represents the stresses on this inclined plane. Again the answers are shown on an element orientated parallel and perpendicular to the**

214

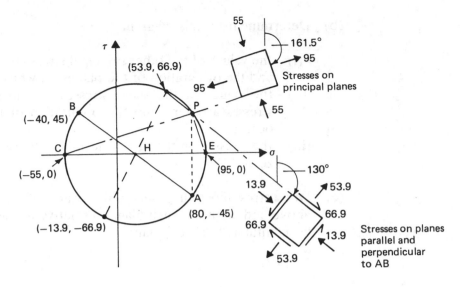

**Figure 9.11**

plane AB. The point on the opposite end of the diameter to this first point represents the stresses on the plane at right angles to the plane AB. These stresses are also shown on the element, as they are required for part (iii) of the question.

$$\text{Normal stress on plane AB} = +53.9 \text{ N/mm}^2$$
$$\underline{\text{Shear stress on plane AB} = +66.9 \text{ N/mm}^2}$$

(iii)

Net strain in direction of AB = direct strain *in direction of AB* $-\, v \times$ direct strain *perpendicular to AB*

$$\therefore \epsilon_{AB} = \frac{\sigma_{AB}}{E} - v \times \frac{\sigma_{(AB+90°)}}{E}$$

$$= \frac{(-13.90)}{210 \times 10^3} - 0.28 \times \frac{53.90}{210 \times 10^3}$$

$$= \underline{-138 \times 10^{-6}}$$

If you are in doubt about this last calculation, look back at Chapter 4. Note also that the question asks for the strain in the direction of AB, so the direct stress in this direction is that stress acting on the plane perpendicular to AB.

## Example 9.3

(a) Draw a Mohr's circle of stress for the conditions at a point in a structural member where the principal compressive stress is 60 N/mm$^2$ and the normal stresses on horizontal and vertical planes are 40 N/mm$^2$ compressive and 120 N/mm$^2$ tensile, respectively, when the shearing stresses in the vertical plane form a clockwise couple.

215

(b)  Determine from this diagram:

  (i)   the values of the other principal stress and the maximum shear stress and the orientation of the planes on which they occur;
  (ii)  the values of the shear stresses on the planes where the normal stresses are zero and the orientation of the planes on which they occur;
  (iii) the stresses on the planes normal to those where the normal stress is zero.

(c)  Draw stress block diagrams to illustrate each of the conditions referred to above and orientate each diagram relative to the original stress block with horizontal and vertical planes.

(Coventry University)

### Solution 9.3

(a)

To draw the circle, first plot point C with coordinates $(-60, 0)$, as this is given as the principal compressive stress. The normal stress acting on the vertical plane ($\sigma_x$) is given as $+120$ N/mm$^2$ and the normal stress on the horizontal plane ($\sigma_y$) is given as $-40$ N/mm$^2$.

The centre of the circle has coordinates $(\{\sigma_x + \sigma_y\}/2, 0) = (\{120 - 40\}/2, 0) = (40, 0)$. Mark this point (H) on the diagram and with compass centred on this point extend the compass to meet C and draw the circle. Mark the point A which represents the stress condition on the vertical plane. This point lies above the $\sigma$ axis, as the question states that the shear stresses on the vertical plane form a clockwise (hence, positive) couple. Similarly, mark the point B representing the stresses on the horizontal planes.

(b), (c)

First, identify the pole point by drawing through A a vertical line to meet the circle at P. The line is vertical, as A represents the stress condition on the vertical plane.

To answer part (i), join P to E and scale the slope of the line and read the value of the maximum principal stress. The stress condition on the principal planes is indicated on the correctly orientated element.

To obtain the maximum shear stresses, scale the value of $\tau$ corresponding to points D and F. Join P to D, to give the inclination of one of the planes of maximum shear stress and measure the inclination of this line. The stress condition on both planes of maximum shear stress is shown on the correctly orientated element.

> Maximum principal stress $= +140$ N/mm$^2$
> Plane of minimum principal stress orientated at $18.4°$
> Maximum shear stress $= \pm 100$ N/mm$^2$
> Plane of maximum positive shear stress orientated at $153.4°$

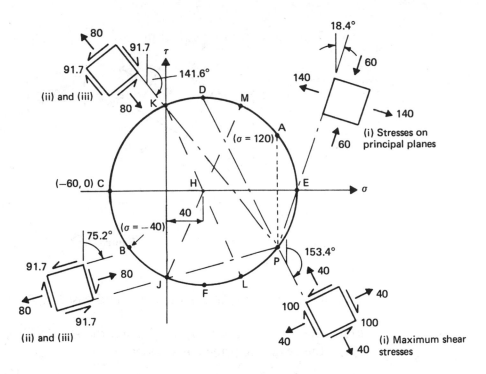

**Figure 9.12**

To answer part (ii), join P to J and K, both points representing the state of stress on planes where the normal stresses are zero. Scale the values of $\tau$ and measure the inclination of these lines. Again these answers are shown on a diagram of a correctly orientated element.

Shear stress for planes of zero normal stress = $\pm 91.7$ N/mm$^2$

Plane indicated by point K orientated at 141.6°

Plane indicated by point J orientated at 75.2°

The stresses on the planes normal to those considered in the answer to part (ii) are given by point L on the opposite end of the diameter to point K and by point M on the opposite end of the diameter to point J. Remember that the points on the opposite end of a diameter represent the state of stress on two mutually perpendicular planes. Scale these stresses from the circle and indicate them on the diagrams of the elements.

Stresses on plane normal to the plane inclined at 141.6°; $\sigma = +80.0$ N/mm$^2$; $\tau = -91.7$ N/mm$^2$
Stresses on plane normal to the plane inclined at 75.2°; $\sigma = +80.0$ N/mm$^2$; $\tau = +91.7$ N/mm$^2$

## Example 9.4

Figures 9.13(a) and (b) show the principal stresses existing at the same point for each of two different applications of load. If these two systems can also occur

217

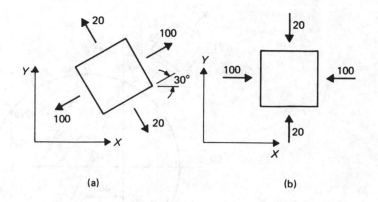

Figure 9.13

simultaneously, obtain for the combined case:

(i)   the direct stresses, $\sigma_x$ and $\sigma_y$ and the shear stress $\tau'$;
(ii)  the resulting principal stresses.

(University of Portsmouth)

### Solution 9.4

(i)
Figure 9.13(a) shows the principal stresses for the first condition of loading. To establish the stresses in the $x$ and $y$ directions, plot the Mohr's circle of stress, using the two known principal stresses as shown in Figure 9.14. To locate the pole point, through E, which represents the maximum principal stress, draw a line

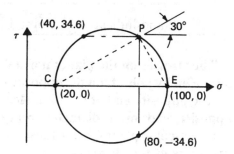

Figure 9.14

parallel to the maximum principal plane to intersect the circle at P. Draw lines through P parallel to the $x$ axis and parallel to the $y$ axis to give the stresses ($\sigma_x$, $\sigma_y$ and $\tau'$) on the two planes parallel to the reference axis. These stresses are plotted on the small stress block in Figure 9.15.

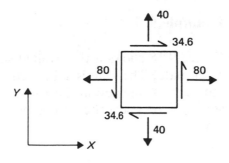

**Figure 9.15**

If the two stress systems can occur simultaneously, then the stresses indicated in Figure 9.15 can be combined with the stress system in Figure 9.13(b) to give the stress system shown in Figure 9.16.

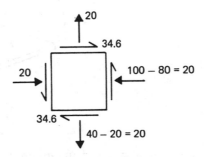

**Figure 9.16**

$$\sigma_x = -20 \text{ N/mm}^2; \ \sigma_y = +20 \text{ N/mm}^2; \ \tau' = \pm 34.6 \text{ N/mm}^2$$

(ii)

The Mohr's circle of stress for the stress system shown in Figure 9.16 can be drawn in the usual way, the pole point found and the principal stresses and the orientation of the principal planes determined as shown in the construction in Figure 9.17.

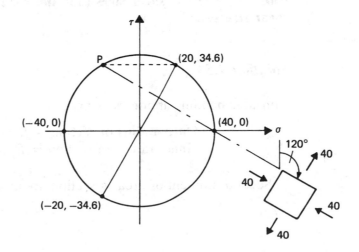

**Figure 9.17**

219

**Example 9.5**

Figure 9.18 shows a thin-walled steel tube welded around its bottom edge to a steel plate. The tube is subjected to a torque $T$, applied about the $y$ axis, and a moment $M$, applied in the $y$–$z$ plane as shown. It is necessary to investigate the state of stress near the weld at point A, which is at the end of the diameter on the $z$ axis.

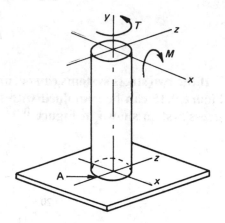

**Figure 9.18**

Using the dimensions given below, find the maximum principal stress and the maximum shear stress in the $x$–$y$ plane in the tube wall at A. Give the angular location of these stresses with respect to the $x$ axis.

The torque $T = 0.25$ kNm and the moment $M = 0.3$ kNm. The mean radius of the tube is 20 mm and the wall thickness is 1 mm.

(Manchester University)

**Note that in this problem the stress condition has to be first calculated from consideration of bending about the $x$ axis, which gives rise to a direct longitudinal tensile stress at A in the direction of the $y$ axis and from consideration of torsion, which gives rise to shear stresses in the $x$–$y$ plane together with complementary shear stresses.**

*Solution 9.5*

Consider bending about the $x$ axis:

Outer radius of section $= R_o = 20 + 0.5 = 20.5$ mm
Inner radius of section $= R_i = 20 - 0.5 = 19.5$ mm

Second moment of area of section about a diameter $= \pi \dfrac{(R_o^4 - R_i^4)}{4}$

$$= \pi \frac{(20.5^4 - 19.5^4)}{4}$$

$$= 25\ 148.43 \text{ mm}^4$$

Bending stress in section: $\sigma = \dfrac{My}{I} = \dfrac{0.3 \times 10^6 \times 20.5}{25\ 148.43}$

$$= 244.55 \text{ N/mm}^2$$

Consider torsion about the longitudinal axis:

For a thin-walled tube,

$$\text{torsional stress} = \frac{T}{2\pi r^2 t} = \frac{0.25 \times 10^6}{2 \times \pi \times 20^2 \times 1}$$

$$= 99.47 \text{ N/mm}^2$$

**This stress condition at A is shown on the small element of material in Figure 9.19(a), for which the Mohr's circle of stress is shown in Figure 9.19(b).**

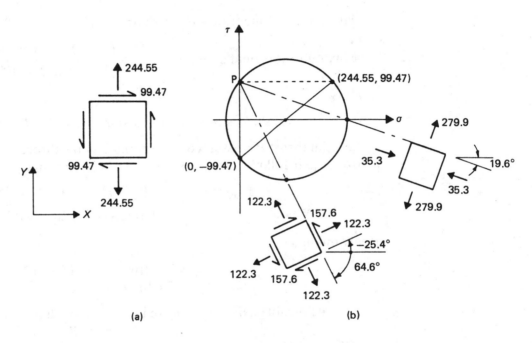

Figure 9.19

The pole point is located in the usual way and the principal stresses and maximum shear stresses are shown on the correctly orientated stress blocks as shown.

Maximum principal stresses = +279.9 N/mm² acting on plane inclined at +19.6°

Maximum shear stress = ± 157.6 N/mm² acting on planes inclined at −25.4° and +64.6°

**Example 9.6**

A drive shaft is designed as a solid circular shaft of 60 mm diameter. It transmits a torsional moment of 2 kNm together with an axial thrust, $P$. If the shear stresses in the shaft are limited to 60 N/mm², calculate the largest value of the thrust $P$ which the shaft can sustain.

(Coventry University)

**This example is included to illustrate the use of one of the formulae given in the Fact Sheet as an alternative solution to drawing the Mohr's circle of stress. The question *can* be solved by drawing the Mohr's circle and the reader should attempt both the algebraic and the graphical solution to this problem. In attempting a graphical solution remember that the diameter of the Mohr's circle of stress is twice the maximum shear stress.**

*Solution 9.6*

Polar second moment of area of shaft $= \frac{1}{2}\pi R^4 = \frac{1}{2}\pi \times 30^4 = 1.272 \times 10^6$ mm⁴

Maximum torsional stress in shaft $= \dfrac{TR}{J} = \dfrac{2 \times 10^6 \times 30}{1.272 \times 10^6} = 47.17$ N/mm²

Thus,

shear stress due to torsion, $\tau' = 47.17$ N/mm²

If the axial thrust $P$ is assumed to be applied in the direction of the $x$ axis, causing a stress $\sigma_x$, it follows that the stress $\sigma_y = 0$ N/mm² and

$$\tau_{max} = 60 = \pm [\{0.5(\sigma_x - \sigma_y)\}^2 + \tau'^2]^{1/2}$$
$$= \pm [\{0.5(\sigma_x)\}^2 + 47.17^2]^{1/2}$$

Rearranging,

$$\sigma_x = 2 \times [(60^2 - 47.17^2)^{1/2}]$$
$$= 74.16 \text{ N/mm}^2$$

∴ Allowable axial thrust in shaft $=$ area of shaft $\times$ axial stress $(\sigma_x)$
$$= \pi \times R^2 \times \sigma_x$$
$$= \pi \times 30^2 \times 74.16 \times 10^{-3}$$
$$= \underline{209.68 \text{ kN}}$$

**Example 9.7**

A steel tube of bore 22 mm and wall thickness 4 mm is subjected to a tensile force of 15 kN and simultaneously a torque of 90 Nm.

(a) Estimate the principal stresses at the outer surface of the tube.
(b) What value of strain would you expect from a strain gauge affixed to the outer surface of the tube with the gauge axis at 45° to the principal axis? (Take $E = 200$ kN/mm², $\nu = 0.3$.)

(University of Aberdeen)

**Solution 9.7**

(a) Consider the tensile axial stress in the section:

$$\text{Cross-sectional area of section} = \pi(R^2 - r^2)$$
$$= \pi(15^2 - 11^2)$$
$$= 326.73 \text{ mm}^2$$

$$\therefore \text{Axial stress in section} = \frac{P}{A} = \frac{15\,000}{326.73} = 45.91 \text{ N/mm}^2$$

Consider the shear stresses due to torsion on the outer surface:

$$\text{Polar second moment of area, } J = \tfrac{1}{2}\pi\,(R^4 - r^4)$$
$$= \tfrac{1}{2}\pi\,(15^4 - 11^4)$$
$$= 56\,523.49 \text{ mm}^4$$

$$\therefore \text{Maximum torsional stress} = \frac{TR}{J} = \frac{90 \times 10^3 \times 15}{56\,523.49}$$
$$= 23.88 \text{ N/mm}^2$$

**The state of stress on the outer surface of the tube is shown in Figure 9.20(a), and, using these stresses, the Mohr's circle of stress is plotted and the principal stresses read off the diagram. Note that the direction of action of the shear stresses on the planes parallel and perpendicular to the reference axes have been assumed to be as shown. If they acted in the reverse direction, this would not have any influence on the magnitude of the principal stresses.**

$$\therefore \text{Maximum principal stress} = +56.1 \text{ N/mm}^2$$

$$\text{Minimum principal stress} = -10.2 \text{ N/mm}^2$$

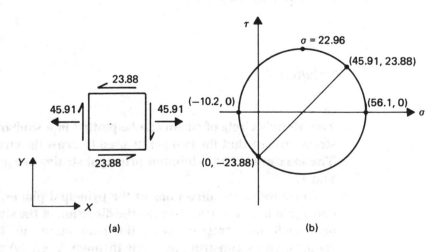

**Figure 9.20**

223

(b)

The strain gauge is affixed with the gauge axis orientated at 45° to the principal planes. Hence, the gauge is aligned in the direction of one of the planes of maximum shear stress and, from the Mohr's circle, the normal stress in this direction is +22.96 N/mm². The plane at right angles to the direction of the gauge axis is also a plane of maximum shear stress on which the normal stress is also +22.96 N/mm².

$$\text{Direct strain in direction of gauge} = \frac{\sigma}{E} = \frac{22.96}{200 \times 10^3} = 0.1148 \times 10^{-3}$$

Similarly,

$$\text{direct strain perpendicular to gauge axis} = 0.1148 \times 10^{-3}$$

Allowing for the Poisson's ratio effect,

net strain in direction of gauge axis = direct strain in this direction
$$\qquad\qquad - \text{Poisson's ratio} \times \text{lateral strain}$$
$$= 0.1148 \times 10^{-3} - (0.3 \times 0.1148 \times 10^{-3})$$
$$= 0.080\,36 \times 10^{-3}$$
$$= 80.36 \times 10^{-6}$$
$$= 80.36 \text{ microstrain}$$

## Example 9.8

At a point in a thin plate the strains are known to be:

$$\epsilon_x = 0.000\,42$$
$$\epsilon_y = 0.000\,18$$
$$\gamma = 0.000\,35$$

(a) Find the magnitude and directions of the principal strains.
(b) Taking $E = 205$ kN/mm² and $\nu = 0.3$, determine the corresponding principal stresses.

(Sheffield University)

## Solution 9.8

(a)

The Mohr's circle of strain can be plotted in a similar way to the Mohr's circle of stress, except that the two points used to draw the circle are $(\epsilon_x, \frac{1}{2}\gamma)$ and $(\epsilon_y, -\frac{1}{2}\gamma)$. The maximum and minimum principal strains are given by the points marked E and C.

To establish the directions of the principal planes, the pole point construction can again be used. However, as the direction of the strain $\epsilon_x$ is normal to the plane on which the stress $\sigma_x$ acts, the pole point for the strain diagram can be established by constructing a line through A $(\epsilon_x, \frac{1}{2}\gamma)$ parallel to the direction of $\epsilon_x$ and not parallel to the plane on which the stress $\sigma_x$ acts. Alternatively, through B

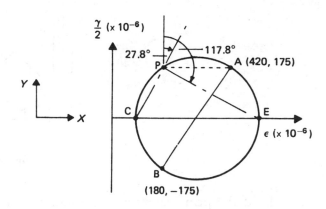

**Figure 9.21**

draw a line parallel to $\epsilon_y$ to meet the circle. To obtain the directions of the principal strains, join P to C and to E. The slope of these lines gives the directions of the principal strains.

Maximum principal strain = $512 \times 10^{-6}$ at 117.8° to direction of $y$ axis
Minimum principal strain = $88 \times 10^{-6}$ at 27.8° to direction of $y$ axis

(b) If the maximum principal strain is $\epsilon_1$, corresponding to a maximum principal stress $\sigma_1$, and the minimum principal strain is $\epsilon_2$, corresponding to a minimum principal stress $\sigma_2$, then

$$\epsilon_1 = \frac{1}{E}(\sigma_1 - \nu\sigma_2)$$

and

$$\epsilon_2 = \frac{1}{E}(-\nu\sigma_1 + \sigma_2)$$

Rearranging the above equations in terms of the principal stresses and substituting for the values of the calculated strains,

$$\sigma_1 = \frac{E(\epsilon_1 + \nu\epsilon_2)}{(1-\nu^2)} = \frac{205 \times 10^3 (512 + 0.3 \times 88)10^{-6}}{(1-0.3^2)} = \underline{121.29 \text{ N/mm}^2}$$

$$\sigma_2 = \frac{E(\epsilon_2 + \nu\epsilon_1)}{(1-\nu^2)} = \frac{205 \times 10^3 (88 + 0.3 \times 512)10^{-6}}{(1-0.3^2)} = \underline{54.43 \text{ N/mm}^2}$$

**Example 9.9**

The strains measured by the three gauges of the 45° rosette shown in Figure 9.22 are:

$$\epsilon_1 = -2 \times 10^{-4}$$
$$\epsilon_2 = +6 \times 10^{-4}$$
$$\epsilon_3 = +4 \times 10^{-4}$$

225

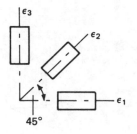

**Figure 9.22**

Find the principal strains and indicate the direction of the maximum principal strain on a rosette diagram.

(University of Westminster)

**To answer this problem it is necessary to remember the formula for the radius and central point of the Mohr's circle of strain which can be plotted from the three measured strains in the 45° rosette. These formulae are available from standard texts and are quoted and used here.**

*Solution 9.9*

The centre of the circle is at

$$\epsilon = \tfrac{1}{2}(\epsilon_1 + \epsilon_3)$$
$$= \tfrac{1}{2}(-2 + 4) \times 10^{-4}$$
$$= 1 \times 10^{-4}$$

The radius of the circle is given by

$$r = \tfrac{1}{2}[(\epsilon_1 - \epsilon_3)^2 + (\epsilon_1 + \epsilon_3 - 2\epsilon_2)^2]^{1/2}$$
$$= \tfrac{1}{2}[(-2 - 4)^2 + (-2 + 4 - \{2 \times 6\})^2]^{1/2} \times 10^{-4}$$
$$= 5.83 \times 10^{-4}$$

**The principal strains can now be calculated from the above information, but as the question also asks for the direction of the principal strains, it is best to construct the Mohr's circle with centre at $1 \times 10^{-4}$ and of radius $5.83 \times 10^{-4}$. The pole point construction can then be used, as in Example 9.8, to determine the direction of the principal strains. This construction is shown in Figure 9.23.**

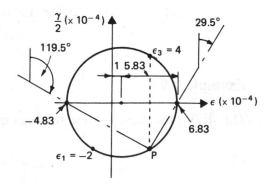

**Figure 9.23**

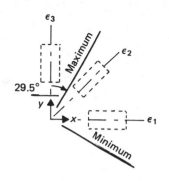

**Figure 9.24**  Direction of principal strains

Maximum principal strain $= 6.8 \times 10^{-4}$ at 29.5° to direction of $y$ axis
Minimum principal strain $= -4.8 \times 10^{-4}$ at 119.5° to direction of $y$ axis

**The formulae used in this example can be used to answer Problem 9.5.**

# 9.5  Problems

**9.1**  The stresses acting on an element of material in a loaded structure are found from strain measurements to be as shown in Figure P9.1

  (i)   Draw the Mohr's circle for stress for the element shown.
  (ii)  Determine the values of the maximum and minimum principal stresses and the maximum shear stress.
  (iii) On a diagram of the element show the positions of the planes on which the principal stresses act.

(Nottingham University)

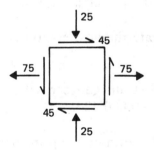

**Figure P9.1**

**9.2**  Figure P9.2 shows the stresses acting on a small element of a structural member.

  (i)   For this small element draw the Mohr's circle of stress.
  (ii)  Hence evaluate the principal stresses and the maximum and minimum shear stresses.
  (iii) Determine the angles between the horizontal and the planes on which there are no normal stresses.

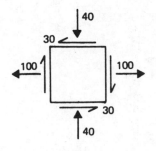

**Figure P9.2**

(iv) Indicate on correctly orientated diagrams the magnitude and sense of the shearing stress on each of the planes found in (iii) above.

(Coventry University)

**9.3** A drill shaft is a hollow steel tube of outside diameter 60 mm and inside diameter 50 mm.

During operation the shaft is subjected to an axial compressive force of 70 kN. Find the maximum torque which may be applied to the shaft if the maximum shear stress must not exceed 80 N/mm². What are the principal stresses in this shaft under this maximum load condition?

(Manchester University)

**9.4** At a point in an elastic material normal stresses of 50 N/mm² tension and 30 N/mm² compression are applied on planes at right angles to one another. If the maximum principal stress is limited to 60 N/mm², determine:

(i) the greatest shearing stress that may be added to the given planes;
(ii) the corresponding principal stresses;
(iii) the inclination of the principal planes;
(iv) the principal strains, assuming $E = 207$ kN/mm² and $\nu = 0.3$.

Illustrate the answers to (ii) and (iii) by means of a clear diagram.

(Nottingham Trent University)

**9.5** A 45° strain gauge rosette gave readings of $\epsilon_0 = 0.000\ 291$, $\epsilon_{45} = 0.000\ 291$, $\epsilon_{90} = -0.000\ 420$.

(i) Determine the principal strains.
(ii) Taking $E = 200 \times 10^3$ N/mm² and $\nu = 0.3$, determine the principal stresses and the maximum shear stress.

(Sheffield University)

# 9.6 Answers to Problems

**9.1** (i) See Figure P9.3; (ii) 92.3 N/mm², −42.3 N/mm², ± 67.3 N/mm²; (iii) see Figure P9.3

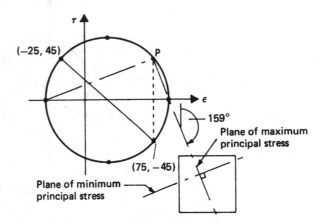

**Figure P9.3**

**9.2** (i) See Figure P9.4; (ii) 106.2 N/mm², −46.2 N/mm², ± 76.2 N/mm²; (iii) 45°, −21.8°; (iv) see Figure P9.4

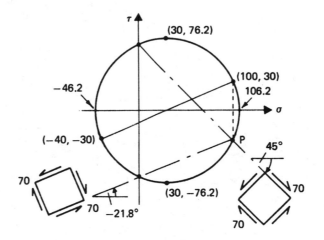

**Figure P9.4**

**9.3** 1.52 kNm, 39.5 N/mm², −120.5 N/mm²

**9.4** (i) $\pm$ 30 N/mm$^2$; (ii) 60 N/mm$^2$, $-40$ N/mm$^2$; (iii) 18.4°, 108.4°; (iv) 0.000 348, $-0.000$ 280

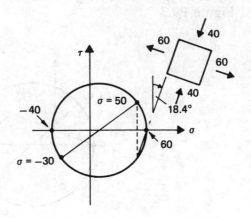

**Figure P9.5**

**9.5** (i) $438 \times 10^{-6}$, $-567 \times 10^{-6}$; (ii) 58.9 N/mm$^2$, $-95.7$ N/mm$^2$, 77.3 N/mm$^2$

# 10 Composite Sections

## 10.1 Contents

Analysis of stress, strain and deformation • Axially loaded composite members • Composite beams in flexure • Stress and strain in composite members due to thermal effects and shrinkage.

## 10.2 The Fact Sheet

### (a) Axially Loaded Members

The analysis of a composite axially loaded member can usually be carried out by applying the principles of *equilibrium* and *compatibility of deformations*. For example, for the axially loaded composite member shown in Figure 10.1 the equations of equilibrium and compatibility are given by

$$P = \sigma_1 A_1 + \sigma_2 A_2$$

and

$$\epsilon_1 = \epsilon_2$$

respectively, where

$$\epsilon_1 = \frac{\sigma_1}{E_1} \quad \text{and} \quad \epsilon_2 = \frac{\sigma_2}{E_2}$$

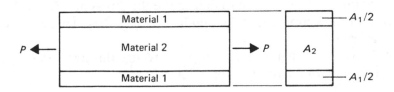

**Figure 10.1**

$A_1$ and $A_2$ are the cross-sectional areas of materials 1 and 2, respectively; $\sigma_1$ and $\sigma_2$ are the stresses; $\epsilon_1$ and $\epsilon_2$ are the strains; and $E_1$ and $E_2$ are the Young's moduli of elasticity.

The equations of equilibrium and compatibility can be combined to give the stress in each of the two materials:

$$\sigma_1 = \frac{P \times E_1}{A_1 E_1 + A_2 E_2}$$

and

$$\sigma_2 = \frac{P \times E_2}{A_1 E_1 + A_2 E_2}$$

A composite section made of more than two materials can be analysed in a similar way. It is assumed in all cases that the axial load causes axial straining without flexural deformation.

## (b) Flexural Members

To analyse a composite beam with a cross-section such as that shown in Figure 10.2(a):

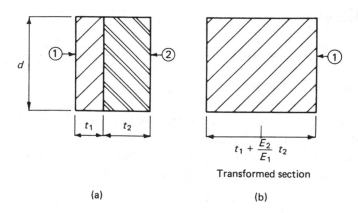

Transformed section

(a)                                      (b)

**Figure 10.2**

(i)   Transform the section into an equivalent section in one material (say material 1) as shown in Figure 10.2(b).

(ii)  Calculate the second moment of area of the transformed section:

$$I_{1(equiv)} = \left[ t_1 + \frac{E_2}{E_1} t_2 \right] \frac{d^3}{12} = t_{1(equiv)} \frac{d^3}{12}$$

(iii) Analyse the transformed section to determine the stress at any level, using the equation of the simple theory of bending ($\sigma = yM/I$). This will give the stress in material 1.

(iv)  To obtain the stress in material 2 at any level where the stress in material 1 is known, use the *compatibility* expression:

$$\sigma_2 = \frac{E_2}{E_1} \sigma_1$$

## (c)  Thermal Stresses

A change of temperature in a composite bar where the materials of construction have different coefficients of expansion will give rise to an internal stress situation within the structure. Most problems of this nature can be analysed by use of principles of *equilibrium* and *compatibility of displacements*. For example, for the composite member shown in Figure 10.3, where the bar and tube are

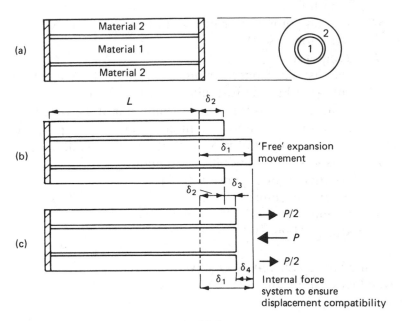

**Figure 10.3**

subjected to a temperature rise $T$, the equations of equilibrium and compatibility are given by

$$P_1 = P_2 = P$$

and

$$\delta_2 + \delta_3 = \delta_1 - \delta_4$$

respectively, where

$$\delta_1 = \alpha_1 LT; \; \delta_2 = \alpha_2 LT; \; \delta_3 = \frac{PL}{A_2 E_2}; \; \delta_4 = \frac{PL}{A_1 E_1}$$

These can be combined to give

$$P = \frac{(\alpha_1 - \alpha_2)TA_1 E_1 A_2 E_2}{(A_1 E_1 + A_2 E_2)}$$

from which

$$\sigma_1 = \frac{P}{A_1}; \; \sigma_2 = \frac{P}{A_2}$$

Most similar problems involving thermal straining or other forms of lack of fit can be solved by use of the principles of equilibrium and compatibility.

# 10.3 Symbols, Units and Sign Conventions

$A$ = cross-sectional area (mm²)
$E$ = Young's modulus of elasticity (kN/mm²)
$I_{CC}$ = the second moment of area of a section about an axis passing through the centroid of the section (mm⁴)
$I_{XX}$ = the second moment of area of a section about an $X$–$X$ axis (mm⁴)
$L$ = length (m)
$M$ = bending moment or moment of resistance (kNm)
$P$ = axial force (kN)
$T$ = temperature (°C)
$y$ = the distance from the neutral axis to the level at which the *bending stress* is calculated in the formula $\sigma = yM/I$ (mm)
$z$ = a distance used for the determination of the position of the neutral axis of a section ($y$ is more commonly used for this purpose but $z$ is used in this text to avoid confusion with the term $y$ as defined above) (mm)
$Z$ = elastic section modulus of a section (mm³)
$\alpha$ = coefficient of expansion (/°C)
$\delta$ = extension (mm)
$\epsilon$ = strain
$\sigma$ = stress (N/mm²)

Tensile stress is taken as positive.
Axial extensions are taken as positive.

# 10.4 Worked Examples

**Example 10.1**

A very rigid beam, 1 m long, is supported in a horizontal position by two short columns made of different materials, one either end of the beam (Figure 10.4). A load of 100 kN is applied to the beam at a point 300 mm from the left-hand end and when supporting this load the beam remains horizontal. If the cross-sectional area of the left-hand column is 500 mm², calculate (a) the cross-sectional area of the right-hand column and (b) the stresses in both columns. Young's moduli of elasticity for the left- and right-hand columns are 200 kN/mm² and 100 kN/mm², respectively.

(Coventry University)

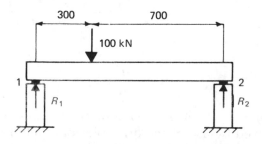

**Figure 10.4**

*Solution 10.1*

(a)  Taking moments about end 1 of the beam,

$(\Sigma M_1 = 0)$

$$(100 \times 300) - (R_2 \times 1000) = 0$$
$$\therefore R_2 = 30 \text{ kN}$$

Resolving vertically,
$(\Sigma V = 0)$

$$R_1 + R_2 = 100$$
$$\therefore R_1 = 100 - 30$$
$$= 70 \text{ kN}$$

As the beam remains horizontal, the change in length of the columns must be the same and the strains must be the same. The *compatibility* equation is, therefore, given by

$$\epsilon_1 = \epsilon_2$$

or

$$\frac{P_1}{A_1 E_1} = \frac{P_2}{A_2 E_2}$$

$$\therefore A_2 = \frac{P_2 A_1 E_1}{P_1 E_2}$$

$$= \frac{30 \times 500 \times 200}{70 \times 100}$$

$$= \underline{428.57 \text{ mm}^2}$$

(b)

$$\sigma_1 = \frac{P_1}{A_1} = \frac{70\ 000}{500} = \underline{140 \text{ N/mm}^2}$$

and

$$\sigma_2 = \frac{P_2}{A_2} = \frac{30\ 000}{428.57} = \underline{70 \text{ N/mm}^2}$$

**Example 10.2**

A composite member ABCD consists of a brass rod BC, 50 mm in diameter and 200 mm long, to the ends of which are concentrically welded steel rods of 20 mm diameter and 100 mm in length. The member is supported at A and D as shown (Figure 10.5). When an axially aligned force is applied at D in the positive $x$ direction, point D moves 0.15 mm. Find:

(a)  the axial movement of points B and C;
(b)  the axial stress in each part of the composite member.

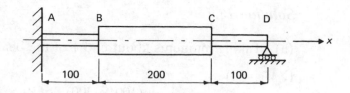

**Figure 10.5**

Ignore self-weight and take $E_{brass} = 120$ kN/mm$^2$, $E_{steel} = 208$ kN/mm$^2$.

(Manchester University)

*Solution 10.2*

(a)

$$\text{Area of steel sections} = A_S = \tfrac{1}{4}\pi d^2 = \tfrac{1}{4}\pi 20^2 = 314.16 \text{ mm}^2$$
$$\text{Area of brass section} = A_B = \tfrac{1}{4}\pi d^2 = \tfrac{1}{4}\pi 50^2 = 1963.49 \text{ mm}^2$$

Let the applied axial force $= P$. The force will be uniform along the length of the composite bar. Hence, the *equilibrium* equation is given by

$$P_B = P_S = P \tag{1}$$

The total extension of the bar is the sum of the extensions of the steel sections and the extension of the brass section. Hence, the *compatibility* expression is

$$\delta_S + \delta_B = 0.15$$

$$\therefore \left[\frac{P_S \times L_S}{E_S A_S}\right] + \left[\frac{P_B \times L_B}{E_B A_B}\right] = 0.15 \tag{2}$$

Hence, combining Equations (1) and (2),

$$P\left[\frac{(2 \times 100)}{208 \times 10^3 \times 314.16} + \frac{200}{120 \times 10^3 \times 1963.49}\right] = 0.15$$

$$\therefore P \times 10^{-6} (3.061 + 0.849) = 0.15$$
$$P = 38\,363.17 \text{ N}$$
$$= 38.36 \text{ kN}$$

Movement of point B = extension of AB (as A is fixed)

$$= \frac{P_S L_{AB}}{A_S E_S} = \frac{38.36 \times 10^3 \times 100}{314.16 \times 208 \times 10^3} = \underline{0.0587 \text{ mm}}$$

Movement of point C = movement of point B + extension of BC

$$= 0.0587 + \frac{P_B L_{BC}}{A_B E_B}$$

$$= 0.0587 + \frac{38.36 \times 10^3 \times 200}{1963.49 \times 120 \times 10^3} = \underline{0.0913 \text{ mm}}$$

(b)

$$\sigma_{AB} = \sigma_{CD} = \frac{P}{A_S} = \frac{38.36 \times 10^3}{314.16} = \underline{122.10 \text{ N/mm}^2}$$

$$\sigma_{BC} = \frac{P}{A_B} = \frac{38.36 \times 10^3}{1963.49} = \underline{19.54 \text{ N/mm}^2}$$

## Example 10.3

Figure 10.6 shows a plan view of a short column consisting of two materials securely bonded together, for which the modular ratio is 20. It carries a point load of 200 kN at the point indicated. Determine the maximum stress in each material.

(University of Westminster)

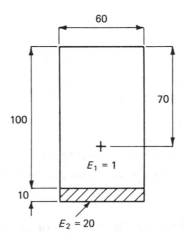

**Figure 10.6**

## *Solution 10.3*

**This is a problem of axial loading combined with bending due to the eccentricity of the load from the section's centroidal axis. To tackle the problem, the section must be transformed to an equivalent section made of one material. Figure 10.7 shows the section transformed to material 1, material 2 being replaced by an equivalent area of material 1 by multiplying its width by the modular ratio. It is necessary to determine the position of the centroid and the second moment of area of the section about the centroid of the section. The calculations are set out in a table.**

| Part | Area ($A$) ($\text{mm}^2 \times 10^3$) | $z$ (mm) | $Az$ ($\text{mm}^3 \times 10^3$) | $I_{CC}(bd^3/12)$ ($\text{mm}^4 \times 10^6$) | $h(=z - \bar{z})$ (mm) | $Ah^2$ ($\text{mm}^4 \times 10^6$) |
|------|------|------|------|------|------|------|
| A | 6 | 60 | 360 | 5.0 | 36.67 | 8.068 |
| B | 12 | 5 | 60 | 0.1 | 18.33 | 4.032 |
| | 18 | | 420 | 5.1 | | 12.100 |

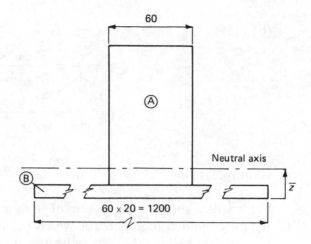

**Figure 10.7** Transformed section

$$\bar{z} = \frac{\Sigma Az}{\Sigma A} = \frac{420 \times 10^3}{18 \times 10^3}$$

$$= 23.33 \text{ mm}$$

$$\begin{array}{r} 12.10 \times 10^6 \\ \underline{5.10 \times 10^6} \end{array}$$

$$I_{XX} = 17.20 \times 10^6 \text{ mm}^4$$

Eccentricity of load from centroidal axis = $e$ = 110 − 70 − 23.33 = 16.67 mm
Moment of load about centroidal axis = $P \times e$ = 200 × 16.67 × $10^{-3}$
$$= 3.334 \text{ kNm.}$$

**Note that the moment of the load about the centroidal axis causes tension in the top face and compression in the bottom face of the column. The maximum stresses will occur at the outer faces of the column.**

Hence,

$$\text{Stress at top face} = -\frac{P}{A} + \frac{My_{max}}{I}$$

$$= -\frac{200 \times 10^3}{18\,000} + \frac{3.334 \times 10^6 \times (110 - 23.33)}{17.20 \times 10^6}$$

$$= -11.11 + 16.80$$

$$= \underline{5.69 \text{ N/mm}^2}$$

$$\text{Stress at bottom face} = -\frac{P}{A} - \frac{My_{max}}{I}$$

$$= -\frac{200 \times 10^3}{18\,000} - \frac{3.334 \times 10^6 \times 23.33}{17.20 \times 10^6}$$

$$= -11.11 - 4.52$$

$$= \underline{-15.63 \text{ N/mm}^2}$$

**This is the stress in the transformed section at the level of the bottom of the column. To obtain the stress in material 2 in the original section at this level, this stress must be multiplied by the modular ratio.**

Hence,

$$\text{maximum stress in material 2} = 20 \times (-15.63)$$
$$= -312.60 \text{ N/mm}^2$$

## Example 10.4

(a) A timber joist having a rectangular cross-section 100 mm wide by 200 mm deep is simply supported over a span of 6 m. Show that the maximum stress in the timber is 13.50 N/mm$^2$ when the beam carries a UDL of 2 kN/m over the complete span.

(b) To strengthen the beam, 100 mm $\times$ 5 mm thick steel plates are securely attached to the side faces of the joist as illustrated in Figure 10.8. These plates extend over only the central 2.5 m of the 6 m span. If the modular ratio $E_s/E_t = 20$, by what percentage may the original UDL of 2 kN/m be increased without the timber stress exceeding the value given in part (a) of the question?

(Nottingham Trent University)

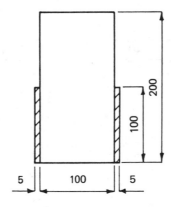

**Figure 10.8**

*Solution 10.4*

(a)

$$\text{Maximum moment at mid-span} = M = wL^2/8 = 2 \times 6^2/8 = 9.0 \text{ kNm}$$
$$\text{Elastic section modulus} = Z = bd^2/6 = 100 \times 200^2/6 = 0.667 \times 10^6 \text{ mm}^3$$

Hence,

$$\text{maximum bending stress} = \frac{M}{Z} = \frac{9.0 \times 10^6}{0.667 \times 10^6} = 13.50 \text{ N/mm}^2$$

(b)
**To answer the second part of the question, it is necessary to transform the section to one material and to determine the neutral axis and the second moment of area of the transformed section about the neutral axis. Figure 10.9 shows the section transformed to an equivalent timber section. The calculations are tabulated below.**

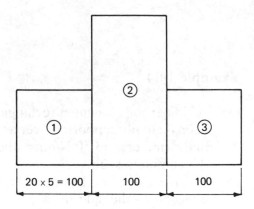

**Figure 10.9**  Transformed section

| Part | Area ($A$) ($mm^2 \times 10^3$) | $z$ (mm) | $Az$ ($mm^3 \times 10^3$) | $I_{CC}(bd^3/12)$ ($mm^4 \times 10^6$) | $h(=z - \bar{z})$ (mm) | $Ah^2$ ($mm^4 \times 10^6$) |
|------|------|------|------|------|------|------|
| 1 | 10 | 50 | 500 | 8.33 | 25 | 6.25 |
| 2 | 20 | 100 | 2000 | 66.67 | 25 | 12.50 |
| 3 | 10 | 50 | 500 | 8.33 | 25 | 6.25 |
|   | 40 |   | 3000 | 83.33 |   | 25.00 |

$$\bar{z} = \frac{\Sigma Az}{\Sigma A} = \frac{3000 \times 10^3}{40 \times 10^3}$$

$$= 75 \text{ mm}$$

$$\frac{25.00 \times 10^6}{83.33 \times 10^6}$$

$$I_{XX} = 108.33 \times 10^6 \text{ mm}^4$$

The maximum bending stress will occur at the furthermost distance from the neutral axis, which in this case is at the top of the beam section. Hence, on the basis of a limiting bending stress of 13.5 N/mm², the moment of resistance of the reinforced section is given by

$$M = \frac{\sigma I}{y_{max}} = \frac{13.5 \times 108.33 \times 10^6 \times 10^{-6}}{(200 - 75)} = 11.70 \text{ kNm}$$

Hence, the maximum load ($w$/unit length) that can be carried by the beam on the basis of the mid-span moment of resistance is given by

$$w = \frac{8M}{L^2} = \frac{8 \times 11.70 \times 10^6}{6000^2} = 2.60 \text{ N/mm} = 2.60 \text{ kN/m}$$

To answer the question completely, it should be realised that there is another critical section at 1.75 m from either support where the steel reinforcing plates are stopped. The moment of resistance of the beam at this section is 9.0 kNm (see answer to part (a)) and the bending moment at this section should be equated to the moment of resistance to obtain the maximum load based on this critical condition.

$$\text{Bending moment at 1.75 m from support} = (\tfrac{1}{2} \times 6 \times w \times 1.75)$$
$$- w \times 1.75 \times (\tfrac{1}{2} \times 1.75)$$
$$= 3.719w \text{ kNm}$$

Hence, equating the bending moment to the moment of resistance,

$$3.718w = 9.0$$
$$\therefore w = 2.42 \text{ kN/m}$$

The allowable maximum load is the smaller of the two calculated values. That is, $w = 2.42$ kN/m.

$$\therefore \text{ percentage increase in load} = \frac{(2.42 - 2.0)}{2.0} \times 100 = \underline{21\%}$$

## Example 10.5

A composite floor construction consists of a concrete floor slab supported on a steel beam as shown in Figure 10.10. The steel beam is propped during casting of the concrete, so that, when the concrete has hardened, the props are removed and all loading is carried by composite beam action.

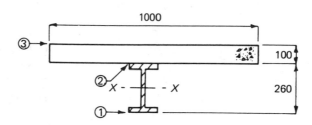

**Figure 10.10**

(a) If the beam spans 10 m and carries a uniformly distributed load of 5 kN/m (including self-weight), determine the maximum stress in the steel at level 1, the maximum stress in the steel and concrete at level 2 and the maximum stress in the concrete at the top of the slab at level 3. Take the modular ratio as 10 and for the steel beam the sectional area = 5500 mm² and $I_{XX} = 65.46 \times 10^6$ mm⁴.

(b) If the stress in the steel is limited to 120 N/mm² in tension or compression and the compressive stress in the concrete is limited to 10 N/mm², calculate the maximum UDL that the beam can carry.

(Coventry University)

241

As in the previous questions, it is first necessary to calculate the position of the neutral axis and the second moment of area of the transformed section about the neutral axis. In problems of this nature it is convenient to transform the section to an equivalent steel section as shown in Figure 10.11. The relevant calculations of the section properties are tabulated below.

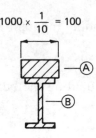

$$1000 \times \frac{1}{10} = 100$$

**Figure 10.11**   Transformed section

## Solution 10.5

| Part | Area ($A$) ($\text{mm}^2 \times 10^3$) | $z$ (mm) | $Az$ ($\text{mm}^3 \times 10^3$) | $I_{CC}$ ($\text{mm}^4 \times 10^6$) | $h(=z - \bar{z})$ (mm) | $Ah^2$ ($\text{mm}^4 \times 10^6$) |
|------|------|------|------|------|------|------|
| A | 10.0 | 310 | 3100 | 8.33 | 63.87 | 40.79 |
| B | 5.5 | 130 | 715 | 65.46 | 116.13 | 74.17 |
| | 15.5 | | 3815 | 73.79 | | 114.96 |

$$\bar{z} = \frac{\Sigma Az}{\Sigma A} = \frac{3815 \times 10^3}{15.5 \times 10^3}$$

$$= 246.13 \text{ mm}$$

$$\begin{array}{c} 114.96 \times 10^6 \\ 73.79 \times 10^6 \\ \hline \end{array}$$

$$I_{XX} = 188.75 \times 10^6 \text{ mm}^4$$

(a)   The maximum bending stresses will occur at the mid-span section of the beam:

maximum bending moment at mid-span, $M = \dfrac{wL^2}{8} = \dfrac{5 \times 10^2}{8} = 62.5 \text{ kNm}$

Hence,

stress in steel at level (1) $= \dfrac{My}{I} = \dfrac{62.5 \times 10^6 \times 246.13}{188.75 \times 10^6} = \underline{81.50 \text{ N/mm}^2}$

stress in steel at level (2) $= \dfrac{My}{I} = \dfrac{62.5 \times 10^6 \times (260.00 - 246.13)}{188.75 \times 10^6}$

$$= \underline{4.59 \text{ N/mm}^2}$$

stress in *concrete* at level (2) = stress in steel at this level $\times \dfrac{E_c}{E_s}$

$$= 4.59 \times \frac{1}{10} = \underline{0.46 \text{ N/mm}^2}$$

stress in steel at level (3) (in transformed section) $= \dfrac{My}{I}$

$$= \frac{62.5 \times 10^6 \times (360.00 - 246.13)}{188.75 \times 10^6} = 37.70 \text{ N/mm}^2$$

stress in *concrete* at level (3) = stress in steel at this level $\times \dfrac{E_c}{E_s}$

$$= 37.70 \times \frac{1}{10} = \underline{3.77 \text{ N/mm}^2}$$

**Note that the modular ratio is given as the ratio of $E_s/E_c$ (= 10), as this is the usual way of describing the modular ratio for a composite steel and concrete problem. To transform the section into an equivalent steel section, it is, therefore, necessary to *divide* the width of the concrete by the modular ratio. Similarly, as the section is transformed to an equivalent *steel* section, it is necessary to *divide* the steel stress in the equivalent section at any level by the modular ratio of 10 to calculate the concrete stress at that level.**

(b)
**The maximum load that the beam can carry depends on whether the concrete reaches its limiting compressive stress of 10 N/mm² before or after the steel reaches its limiting stress of 120 N/mm². By inspection of the calculations in part (a), the maximum stress in the steel occurs at level (1) and the maximum compressive stress in the concrete at level (3).**

Consider the maximum steel stress at level (1). A bending moment of 62.5 kNm causes a steel stress of 81.50 N/mm². Hence, if the bending moment that causes a stress of 120 N/mm² is $M$, then, by simple proportions (as moments and stresses are linearly related),

$$M = \frac{120 \times 62.5}{81.5} = 92.02 \text{ kNm}$$

Similarly, the moment that will cause a compressive stress of 10 N/mm² in the concrete is given by

$$M = \frac{10 \times 62.5}{3.77} = 165.78 \text{ kNm}$$

These two calculations show that the stress in the steel is critical and, hence, the maximum bending moment that the beam can resist is the lower of the two calculated moments—that is, 92.02 kNm.

Hence, the maximum UDL that can be carried is given by

$$w = \frac{8 \times M}{L^2} = \frac{8 \times 92.02}{10^2}$$

$$= \underline{7.36 \text{ kN/m}}$$

**Example 10.6**

Two short concrete columns, 800 mm long and 200 mm $\times$ 200 mm square, are cast from the same batch, one without any reinforcing, and the other with four 20 mm diameter steel bars symmetrically placed as shown in Figure 10.12. After a period of time the unreinforced column is observed to have shortened by 0.16 mm.

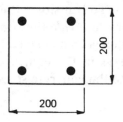

**Figure 10.12**

By what amount would the other column have shortened and what stresses would have developed in the steel and concrete as a result?

Assume that the length of the reinforcing bars is 800 mm, the same as the column, and take $E_s = 210$ kN/mm$^2$ and $E_c = 14$ kN/mm$^2$.

(University of Hertfordshire)

*Solution 10.6*

Area of steel $= A_s = 4 \times (\frac{1}{4}\pi \times d^2) = 4 \times (\frac{1}{4} \times \pi \times 20^2) = 1256.64$ mm$^2$
Area of concrete $=$ gross area of column $-$ area of steel
$$= 200 \times 200 - 1256.64$$
$$= 38\ 743.36 \text{ mm}^2$$

**Figure 10.13 shows both the *unreinforced* column, which shortens by 0.16 mm, and the *reinforced* column, in which the tendency of the concrete to shrink is resisted by the reinforcement. Hence, a compressive force ($P_s$) develops in the steel reinforcement and an equal and opposite tensile force ($P_c$) develops in the concrete. The final length of the reinforced column will, therefore, lie somewhere between the original length and the length of the shortened unreinforced column, as shown.**

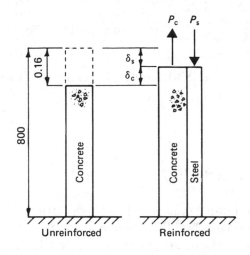

**Figure 10.13**

The *equilibrium* equation for the reinforced section is given by

$$P_c = P_s = P \tag{1}$$

The *compatibility* equation for the reinforced section is given by

$$\delta_c + \delta_s = 0.16 \text{ mm} \tag{2}$$

where

$$\delta_s = \frac{P_s L_s}{A_s E_s} = \frac{PL}{A_s E_s} \tag{3}$$

and

$$\delta_c = \frac{P_c L_c}{A_c E_c} = \frac{PL}{A_c E_c} \tag{4}$$

Hence, combining Equations (2)–(4),

$$\frac{PL}{A_c E_c} + \frac{PL}{A_s E_s} = 0.16$$

$$\therefore \frac{P \times 800}{(38\,743.36 \times 14 \times 10^3)} + \frac{P \times 800}{(1256.64 \times 210 \times 10^3)} = 0.16$$

$$\therefore P(1.475 + 3.032) \times 10^{-6} = 0.16$$
$$P = 35.50 \text{ kN}$$

Hence,

$$\text{stress in steel} = \sigma_s = \frac{P}{A_s} = \frac{35.50 \times 10^3}{1256.64} = \underline{28.25 \text{ N/mm}^2}$$

245

$$\text{stress in concrete} = \sigma_c = \frac{P}{A_c} = \frac{35.50 \times 10^3}{38\,743.36} = \underline{0.92 \text{ N/mm}^2}$$

$$\text{Shortening of column} = \text{shortening of steel} = \delta_s = \frac{PL}{A_s E_s}$$

$$= \frac{35.50 \times 10^3 \times 800}{1256.64 \times 210 \times 10^3} = \underline{0.108 \text{ mm}}$$

**Example 10.7**

A heat exchanger is formed from a series of tubes of a copper/nickel alloy carrying hot fluid surrounded by a water cooling jacket in a steel casing. The dimensions are shown in Figure 10.14. The tubes (36 in total) and casing are built into heavy end manifolds effectively preventing differential longitudinal movement. If the whole assembly is unstressed at 18 °C, determine the longitudinal stresses occurring in tubes and casing when the tubes carry fluid at an average temperature of 120 °C and the casing temperature is 18 °C.

$$E_{\text{steel}} = 210 \text{ kN/mm}^2; \ E_{\text{alloy}} = 180 \text{ kN/mm}^2$$
$$\alpha_{\text{steel}} = 16 \times 10^{-6}/°C; \ \alpha_{\text{alloy}} = 14 \times 10^{-6}/°C$$

(Nottingham University)

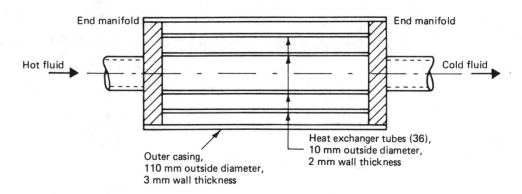

**Figure 10.14**

*Solution 10.7*

Area alloy $= A_a = 36 \times \frac{1}{4} \times \pi \times (D^2 - d^2) = 36 \times \frac{1}{4} \times \pi \times (10^2 - 6^2)$
$\qquad = 1809.56 \text{ mm}^2$
Area steel $= A_s = \frac{1}{4} \times \pi \times (D^2 - d^2) = \frac{1}{4} \times \pi \times (110^2 - 104^2)$
$\qquad = 1008.45 \text{ mm}^2$

**Figure 10.15(a) shows the system where one manifold is released to permit free expansion of the alloy as the temperature rises from 18 °C to 120 °C. As the alloy tubes are attached to the casing, the free expansion is prevented by the casing, which induces a restraining (compressive) force ($P_a$) in the alloy and an equal and opposite tensile force ($P_s$) in the casing. The final position of both tubes and casing is shown in Figure 10.15(b), where the end of the system lies somewhere between**

the original position and the position that would be assumed by the tubes if unrestrained.

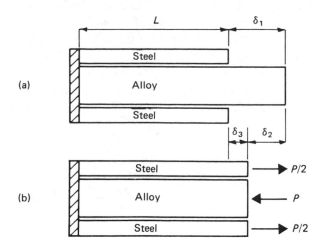

**Figure 10.15**

The *equilibrium* equation for the system is

$$P_s = P_a = P \qquad (1)$$

The *compatibility* equation is given by

$$\delta_1 = \delta_2 + \delta_3 \qquad (2)$$

where

$$\delta_1 = (\alpha L T)_a = 14 \times 10^{-6} \times L \times (120 - 18) = 1428 \times 10^{-6} \times L \qquad (3)$$

$$\delta_2 = \frac{P_a L}{A_a E_a} = \frac{PL}{1809.56 \times 180 \times 10^3} = 0.003\ 07 \times 10^{-6} \times P \times L \qquad (4)$$

$$\delta_3 = \frac{P_s L}{A_s E_s} = \frac{PL}{1008.45 \times 210 \times 10^3} = 0.004\ 72 \times 10^{-6} \times P \times L \qquad (5)$$

Hence, combining Equations (2)–(5),

$$1428 \times 10^{-6} \times L = 0.003\ 07 \times 10^{-6} \times P \times L + 0.004\ 72 \times 10^{-6} \times P \times L$$
$$\therefore P = 183\ 311.94 \text{ N}$$
$$= 183.31 \text{ kN}$$

Hence,

$$\text{steel stress} = \sigma_s = \frac{P}{A_s} = \frac{183.31 \times 10^3}{1008.45} = \underline{181.77 \text{ N/mm}^2}$$

$$\text{alloy stress} = \sigma_a = \frac{P}{A_a} = \frac{183.31 \times 10^3}{1809.56} = \underline{101.30 \text{ N/mm}^2}$$

# 10.5  Problems

**10.1** In Figure P10.1 a load of 50 kN is attached to a rigid beam BDF which is attached to bars AB, CD and EF; each 2 m long. Bars AB and EF are of steel, 12 mm diameter, and bar CD is of aluminium, 15 mm diameter. Find the stress in each bar.

It is now required to increase the applied load to 75 kN without increasing the stresses in bars AB and EF. To do this, the bar CD is shortened by an amount $\delta$. Calculate the value of $\delta$ and the new stress in the bar CD. Take $E_{steel} = 207 \times 10^3$ N/mm$^2$, $E_{aluminium} = 73 \times 10^3$ N/mm$^2$.

(Salford University)

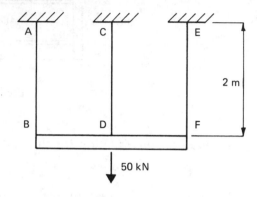

**Figure P10.1**

**10.2** A rectangular section composite tension member 2.5 m long consists of a 100 mm × 50 mm × 2.5 m long timber member fixed firmly between two steel plates 100 mm × 5 mm × 2.5 m long as shown in Figure P10.2.

Find the maximum tensile force which may be applied to the member such that the stress does not exceed 6 N/mm$^2$ in the timber and 125 N/mm$^2$ in the steel. What will be the increase in length of the composite member under the action of this maximum tensile force?

Take $E_{steel} = 200$ kN/mm$^2$; $E_{timber} = 12$ kN/mm$^2$.

(University of Westminster)

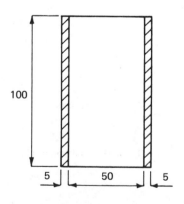

**Figure P10.2**

**10.3**  A reinforced concrete column has a cross-section 250 mm × 250 mm and carries an axial compressive load of 750 kN. The column, which contains 3% of steel in the form of vertical bars, was constructed in the summer, when the temperature was 30 °C.

Calculate the stresses in the steel and the concrete during a winter when the temperature drops to −5 °C. Ignore the loss of concrete cross-sectional area due to the presence of the steel.

For steel: $E_s = 200$ kN/mm$^2$  and $\alpha_s = 10.8 \times 10^{-6}/°C$
For concrete: $E_c = 30$ kN/mm$^2$  and $\alpha_c = 12.6 \times 10^{-6}/°C$

(Manchester University)

**10.4**  Two steel rods, each of cross-sectional area 500 mm$^2$, and an aluminium alloy rod, of cross sectional area 600 mm$^2$, are cut to a length of 3 m and attached to rigid end plates to form a composite member.

If a tensile force of 100 kN is applied to the member and at the same time the temperature of the system is lowered by 20 °C, determine the force developed in each of the rods and the extension of the member.

What further extension of the member will take place if the aluminium alloy is cut?

$E_{steel} = 200$ kN/mm$^2$; $\alpha_{steel} = 12 \times 10^{-6}$ per °C
$E_{alloy} = 80$ kN/mm$^2$; $\alpha_{alloy} = 22 \times 10^{-6}$ per °C

(Nottingham Trent University)

**10.5**  A timber beam with a 150 mm × 150 mm section has a steel plate 150 mm × 12 mm fixed to it as shown in Figure P10.3, to form a composite beam. Calculate the maximum point load that can be carried centrally on a simply supported span of 3 m.

Allowable stress for timber = 8 N/mm$^2$ and for steel = 180 N/mm$^2$.
$E_{timber} = 10$ kN/mm$^2$; $E_{steel} = 200$ kN/mm$^2$

(University of Hertfordshire)

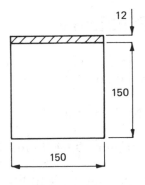

12

150

150

**Figure P10.3**

**10.6**  A steel beam acts compositely with a concrete slab as shown in Figure P10.4. The Young's moduli are

$$E_{conc} = 25 \text{ kN/mm}^2; \ E_{steel} = 205 \text{ kN/mm}^2$$

and the properties of the steel universal beam are

$$A = 28\,840 \text{ mm}^2; \ I_{XX} = 3391.3 \times 10^6 \text{ mm}^4$$

(a) Show that the neutral axis of the combined section is at a height of 681.9 mm above the base.

(b) If the steel stress is limited to 210 N/mm² (in tension) and the concrete stress is limited to 11 N/mm² (in compression), determine the moment of resistance ($M_R$) of the composite section.

(c) Show on a diagram the stress distribution over the composite section when the moment ($M_R$) is being carried.

(Nottingham University)

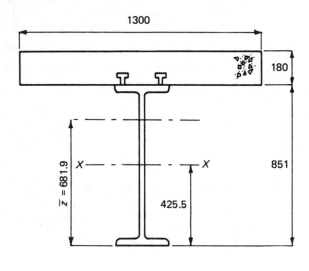

**Figure P10.4**

## 10.6 Answers to Problems

**10.1** $\sigma_{AB} = \sigma_{EF} = 173.25 \text{ N/mm}^2; \ \sigma_{CD} = 61.13 \text{ N/mm}^2; \ \delta = 3.876 \text{ mm};$
$\sigma_{CD} = 202.60 \text{ N/mm}^2$

**10.2** 130 kN; 1.25 mm

**10.3** 77.17 N/mm²; 9.69 N/mm² (both stresses in compression)

**10.4** 72.91 kN (steel); 27.10 kN (aluminium) (both in tension); 0.374 mm; 0.406 mm

**10.5** 11.40 kN

**10.6** (b) 1881.10 kNm; (c) see Figure P10.5

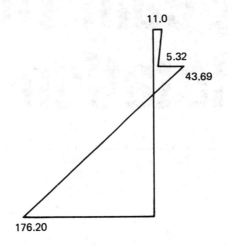

11.0

5.32

43.69

176.20

**Figure P10.5**

# 11 Beam Deflections and Rotations

## 11.1 Contents

Calculation of the deflections and rotations of loaded beams using Macaulay's method of successive integration • Treatment of concentrated loads, uniformly distributed loads covering all or part of the span and couples • Calculation of the position and magnitude of the maximum deflection • Beams simply supported at either end, simply supported beams with cantilevered ends, cantilevered beams.

There are many methods for calculating the deflections and rotations in beams. At first-year level the number of methods considered is usually restricted to one or two and almost certainly the first method considered will be the method of successive integration. This technique, popularly known as Macaulay's method, will be used in this chapter to solve a variety of beam problems.

## 11.2 The Fact Sheet

### (a) Relationship between Bending Moment and Beam Curvature

The relationship between the bending moment at any section along a beam and the curvature at that section is given by

$$M = -EI\frac{1}{R} = -EI\frac{d^2v}{dx^2}$$

or

$$\frac{d^2v}{dx^2} = -\frac{M}{EI} \tag{11.1}$$

where,

$EI$ = the beam's flexural rigidity;
$M$ = the bending moment;
$R$ = the radius of curvature;
$v$ = the vertical deflection; and
$x$ = distance along the beam measured from a chosen origin.

## (b) Calculation of Rotation and Deflection by Successive Integration

The rotation ($dv/dx$) and vertical deflection ($v$) can be determined by integrating equation (11.1) to give the following expressions:

$$\frac{dv}{dx} = \frac{1}{EI} \left[ \int - M \, dx + A \right] \tag{11.2}$$

and

$$v = \frac{1}{EI} \left[ \int\int - M \, dx \, dx + Ax + B \right] \tag{11.3}$$

In the above equations it is assumed that the flexural rigidity ($EI$) is a constant along the length of the beam. If $EI$ varies with distance $x$, then it must be included within the integration on the right-hand side of the equations.

The constants of integration ($A$ and $B$) are found by applying the equations at two locations where the deflection and/or rotation are known. Usually (but not always) these locations will be the supports of the beam. Once $A$ and $B$ are evaluated, then appropriate values of distance $x$ can be inserted into the equations to determine the deflection and rotation at any location along the beam's span.

## (c) Macaulay's Variation

Equations (11.2) and (11.3) are only valid in the region of a beam where a single expression describes the variation of bending moment over that region. Where there are discontinuities in the bending moment diagram, such as at the point of application of a point load, then the equation for bending moment must be written down using 'Macaulay brackets' and the following procedure:

  (i)    Take the origin at the left-hand end of the beam.
  (ii)   Write down the bending moment expression at a section at the extreme right-hand end of the beam in terms of all loads to the left of that section.
  (iii)  Do *not* simplify the bending moment expression by expanding any of the terms involving the distance $x$ which are contained within brackets (Macaulay brackets).
  (iv)   Integrate, using Equations (11.2) and (11.3), *keeping all the bracketed terms within brackets*.
  (v)    Apply the resulting equations to any part of the beam but *neglect all terms within the brackets wherever they are negative or zero*.

### (d)  Treatment of Distributed Loads

Where a distributed load does not extend to the right-hand end of the beam, as shown in Figure 11.1, then introduce two equal but opposite dummy loads, as shown in Figure 11.2. These two loads will have no net effect on the behaviour of the beam. The loading system shown in Figure 11.2 should then be used when deriving the initial bending moment expression.

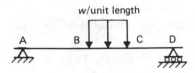

**Figure 11.1**

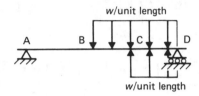

**Figure 11.2**

### (e)  Maximum Deflection

The position of the point where maximum deflection occurs can be established from the fact that the slope of the beam will be zero at this point. Hence, Equation (11.2) can be equated to zero to establish the value of $x$ at which the deflection is a maximum. This value of $x$ can then be substituted into Equation (11.3) to give the value of the maximum deflection.

# 11.3   Symbols, Units and Sign Conventions

$$A, B = \text{constants of integration}$$
$$EI = \text{the beam's flexural rigidity (kNm}^2\text{)}$$
$$M = \text{bending moment (kNm)}$$
$$v = \text{the vertical deflection (mm)}$$
$$dv/dx = \theta = \text{rotation (rad)}$$
$$x = \text{distance measured along the beam (m)}$$

Sagging bending moments are taken as positive.
Downward deflections are taken as positive.
Clockwise rotations are positive.
$x$ is measured from the origin at the left-hand end of the beam in a positive direction towards the right-hand end of the beam.

## 11.4  Worked Examples

### Example 11.1

A horizontal beam with a symmetrical uniform cross-section spans 8 m and is simply supported at its ends. Point loads of 15 kN and 20 kN are applied at 1.5 m and 3.5 m, respectively, from one end. Both loads act vertically downwards.

Calculate a suitable second moment of area for the beam's cross-section, given that the maximum deflection at any point must not exceed 1/400 of the span. Take $E = 200$ kN/mm$^2$.

(Manchester University)

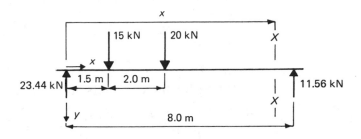

**Figure 11.3**

To start the question, it is necessary to calculate the support reactions. The correct values of the support reactions are shown in Figure 11.3. Check these values before proceeding.

### Solution 11.1

The bending moment at a section $X$–$X$ close to the right-hand end of the span is given by

$$M = 23.44x - 15\{x - 1.5\} - 20\{x - 3.5\}$$

Note the use of the curled $\{\}$ Macaulay brackets, which must *not* be expanded to simplify the moment equation.

Hence, the general equation is given by

$$\frac{d^2v}{dx^2} = -\frac{1}{EI}\left[23.44x - 15\{x - 1.5\} - 20\{x - 3.5\}\right]$$

Integrate once:

$$\frac{dv}{dx} = -\frac{1}{EI}\left[\frac{23.44x^2}{2} - \frac{15\{x - 1.5\}^2}{2} - \frac{20\{x - 3.5\}^2}{2} + A\right]$$

255

Integrate again:

$$v = -\frac{1}{EI}\left[\frac{23.44x^3}{6} - \frac{15\{x - 1.5\}^3}{6} - \frac{20\{x - 3.5\}^3}{6} + Ax + B\right]$$

Boundary condition $x = 0$, $v = 0$ gives $B = 0$.

**Note that this boundary condition comes from the left-hand end support, where there is no vertical deflection. When $x = 0$ is substituted into the final equation, all bracketed terms are negative and therefore are ignored. The second boundary condition comes from consideration of the right-hand support, where again there is no vertical deflection.**

Boundary condition $x = 8$ m, $v = 0$ gives

$$0 = -\frac{1}{EI}\left[\frac{23.44 \times 8^3}{6} - \frac{15\{8 - 1.5\}^3}{6} - \frac{20\{8 - 3.5\}^3}{6} + A \times 8\right]$$

which solves to give $A = -126.24$.

Substituting for the constants $A$ and $B$ gives

$$\frac{dv}{dx} = -\frac{1}{EI}\left[\frac{23.44x^2}{2} - \frac{15\{x - 1.5\}^2}{2} - \frac{20\{x - 3.5\}^2}{2} - 126.24\right] \quad (1)$$

and

$$v = -\frac{1}{EI}\left[\frac{23.44x^3}{6} - \frac{15\{x - 1.5\}^3}{6} - \frac{20\{x - 3.5\}^3}{6} - 126.24 \times x\right] \quad (2)$$

To calculate the position of maximum deflection, equate the equation for the slope (Equation 1) to zero and solve. That is,

$$-\frac{1}{EI}\left[\frac{23.44x^2}{2} - \frac{15\{x - 1.5\}^2}{2} - \frac{20\{x - 3.5\}^2}{2} - 126.24\right] = 0$$

**By inspection of the problem, the point of maximum deflection will be near to mid-span to the right of the 20 kN load. Hence, the solution will be for some value of $x$ greater than 3.5 m and all the terms in the Macaulay brackets will be positive. Hence, all the brackets can be expanded and the equation solved for $x$. If it were incorrectly assumed that the point of maximum deflection is to the left of the 20 kN load, then the solution would give an answer for $x$ which is outside of the region where it is assumed that the maximum deflection occurs and would, hence, be seen to be invalid.**

$$11.72x^2 - 7.5(x^2 - 3 \times x + 2.25) - 10(x^2 - 7 \times x + 12.25) - 126.24 = 0$$
$$\therefore x^2 - 16.0 \times x + 45.95 = 0$$

which has the solution $x = 3.75$ m. Hence, substituting $x = 3.75$ into the equation for deflection (Equation 2),

$$v = -\frac{1}{EI}\left[\frac{23.44 \times 3.75^3}{6} - \frac{15\{3.75 - 1.5\}^3}{6} - \frac{20\{3.75 - 3.5\}^3}{6} - 126.24 \times 3.75\right]$$

$$= \frac{295.91}{EI}\text{ m}$$

Note that in the above equation $v$ will be in metres, provided that $EI$ is expressed in $\text{kNm}^2$.

The deflection is limited to 1/400 of the span. Hence,

$$\frac{295.91}{EI} < \frac{1}{400} \times 8$$

or

$$I > \frac{295.91 \times 400}{8 \times E}$$

$$> \frac{295.91 \times 400}{8 \times (200 \times 10^6)}$$

$$> 73.98 \times 10^{-6} \text{ m}^4$$

$$\underline{> 73.98 \times 10^6 \text{ mm}^4}$$

## Example 11.2

Figure 11.4 shows a uniform horizontal cantilever beam ABC of composite construction, built-in at A and carrying a 5 kN load at the point B, halfway along its length. The cross-section is as shown. The flange plates (Young's modulus, $E_1 = 200 \text{ kN/mm}^2$) are bonded to the core material ($E_2 = 10 \text{ kN/mm}^2$).

(a)  Find the deflection of the point C due to the applied load.
(b)  If additionally the cantilever were to be simply supported at the free end, with C on the same level as A, what would be the reaction at C?

(Cambridge University)

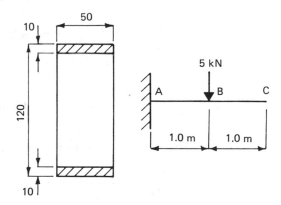

Figure 11.4

## Solution 11.2

(a)
**Figure 11.5 shows the composite beam transformed into an equivalent beam made of material 2. All the subsequent calculations can be carried out using the**

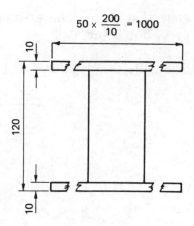

$$50 \times \frac{200}{10} = 1000$$

**Figure 11.5**

properties of this equivalent section. The second moment of area of the equivalent section about the neutral axis can be determined by use of the parallel axis theorem:

$$I = I_{\text{core}} + I_{\text{flanges}}$$

$$= \frac{1}{12} 50 \times 100^3 + 2 \left[ \frac{1}{12} 1000 \times 10^3 + (1000 \times 10) \times 55^2 \right]$$

$$= 64.83 \times 10^6 \text{ mm}^4$$
$$= 64.83 \times 10^{-6} \text{ m}^4$$

The support reactions are shown in Figure 11.6. The bending moment at a section $X$–$X$ close to the right-hand end of the span is given by

$$M = -5 + 5x - 5\{x - 1\}$$

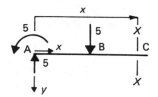

**Figure 11.6**

Hence, the general equation is given by

$$\frac{d^2v}{dx^2} = -\frac{1}{EI}[-5 + 5x - 5\{x - 1\}]$$

Integrate once:

$$\frac{dv}{dx} = -\frac{1}{EI}\left[ -5x + \frac{5x^2}{2} - \frac{5\{x-1\}^2}{2} + A \right] \tag{1}$$

Integrate again:

$$v = -\frac{1}{EI}\left[ -\frac{5x^2}{2} + \frac{5x^3}{6} - \frac{5\{x-1\}^3}{6} + Ax + B \right] \tag{2}$$

Boundary conditions $x = 0$, $v = 0$ give $B = 0$ from Equation (2) (no deflection at support A); $x = 0$, $dv/dx = 0$ give $A = 0$ from Equation (1) (no rotation at support A).

**Note that in this problem the constants of integration are obtained by substituting into both the equation for deflection and the equation for rotation.**

Substituting for the constants $A$ and $B$,

$$v = -\frac{1}{EI}\left[ -\frac{5x^2}{2} + \frac{5x^3}{6} - \frac{5\{x - 1\}^3}{6} \right] \tag{3}$$

The deflection at C can be obtained by substituting $x = 2$ m into Equation (3):

$$v_C = -\frac{1}{EI}\left[ -\frac{5 \times 2^2}{2} + \frac{5 \times 2^3}{6} - \frac{5\{2 - 1\}^3}{6} \right]$$

$$= \frac{4.17}{EI}$$

$$= \frac{4.17}{(10 \times 10^6) \times (64.83 \times 10^{-6})}$$

$$= 6.43 \times 10^{-3} \text{ m}$$

$$= \underline{6.43 \text{ mm}}$$

(b)
**Figure 11.7 shows that the cantilever supported at end C is equivalent to the case considered in part (a) of this question superimposed on to the case where the cantilever is subjected to a single vertically upwards end force $R$ equal in magnitude to the propping reaction.**

For the cantilever considered in part (a), deflection at C = $4.17/EI$. For the cantilever with end loading $R$,

$$\text{deflection at C} = \frac{RL^3}{3EI}$$

$$= \frac{R \times 2^3}{3EI}$$

$$= R \times 2.67/EI$$

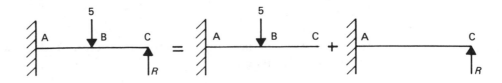

**Figure 11.7**

259

This last formula for the deflection at the end of a cantilever of length $L$ and subjected to a concentrated end load is a standard formula. As such it would be reasonable to expect that it is remembered and quoted in answering this question. If it can not be remembered, it can be derived in a similar way to the answer to part (a).

As the net end deflection at the end of the propped cantilever must be zero, it follows that

$$\frac{R \times 2.67}{EI} = \frac{4.17}{EI}$$

or

$$R = 4.17/2.67 \text{ kN}$$
$$= 1.56 \text{ kN}$$

## Example 11.3

Figure 11.8 shows a timber beam which is to be designed to carry a maximum loading of 5 kN/m and a minimum loading of 1 kN/m. To calculate the maximum mid-span deflection the beam is loaded as shown. Calculate the mid-span deflection in terms of the flexural stiffness ($EI$) of the beam section.

(Coventry University)

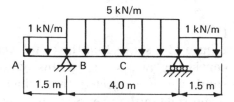

**Figure 11.8**

Use can be made of the symmetry of the problem. As, at the mid-span point C, the rotation of the beam will be zero, the problem can be solved by considering an equivalent beam as shown in Figure 11.9. In this equivalent beam a roller support is introduced at C which will permit vertical deflection but no rotation. The loading can be separated into a UDL of 1 kN/m and a UDL of 4 kN/m, as shown, so that both loads extend to the right-hand end of the beam and thus it is not necessary to introduce any dummy loads.

## Solution 11.3

In Figure 11.9 the bending moment at a section $X$–$X$ close to the right-hand end of the span is given by

$$M = -\frac{1x^2}{2} + 11.5\{x - 1.5\} - \frac{4\{x - 1.5\}^2}{2}$$

260

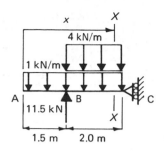

**Figure 11.9**

Hence, the general equation is given by

$$\frac{d^2v}{dx^2} = -\frac{1}{EI}\left[-\frac{1x^2}{2} + 11.5\{x - 1.5\} - \frac{4\{x - 1.5\}^2}{2}\right]$$

Integrate once:

$$\frac{dv}{dx} = -\frac{1}{EI}\left[-\frac{x^3}{6} + \frac{11.5\{x - 1.5\}^2}{2} - \frac{4\{x - 1.5\}^3}{6} + A\right] \quad (1)$$

Integrate again:

$$v = -\frac{1}{EI}\left[-\frac{x^4}{24} + \frac{11.5\{x - 1.5\}^3}{6} - \frac{4\{x - 1.5\}^4}{24} + Ax + B\right] \quad (2)$$

Boundary conditions $x = 3.5$, $dv/dx = 0$ give $A = -10.52$ from Equation (1) (no rotation at C); $x = 1.5$, $v = 0$ give $B = 15.99$ from Equation (2) (no deflection at support B).

Substituting for the constants $A$ and $B$ gives

$$v = -\frac{1}{EI}\left[-\frac{x^4}{24} + \frac{11.5\{x - 1.5\}^3}{6} - \frac{4\{x - 1.5\}^4}{24} - 10.52 \times x + 15.99\right]$$

The mid-span deflection occurs at C, where $x = 3.5$ m. Hence, substituting $x = 3.5$ into the last equation,

$$v = -\frac{1}{EI}\left[-\frac{3.5^4}{24} + \frac{11.5\{3.5 - 1.5\}^3}{6} - \frac{4\{3.5 - 1.5\}^4}{24} - 10.52 \times 3.5 + 15.99\right]$$

which solves to give

$$v = \underline{14.42/EI}$$

## Example 11.4

The uniform simply supported beam ABC is subjected to the loading shown in Figure 11.10. The flexural rigidity of the beam is $EI$. Determine the slope of the beam at A and C and the deflection at B.

(Salford University)

**Again the first stage in the calculations is to determine the support reaction at A. This has been done and marked on the diagram in Figure 11.11. Check the**

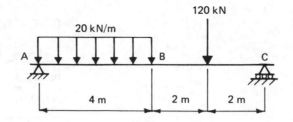

Figure 11.10

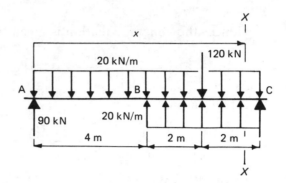

Figure 11.11

value of the reaction before proceeding. This is an example where the uniformly distributed load does not extend to the right-hand end of the span. It is, therefore, necessary to introduce two equal and opposite 'dummy' uniformly distributed loads, as shown in Figure 11.11. It is this second figure which will form the basis of the analysis.

### Solution 11.4

The bending moment at a section $X–X$ close to the right-hand end of the span is given by

$$M = 90x - \frac{20x^2}{2} + \frac{20\{x - 4\}^2}{2} - 120\{x - 6\}$$

Hence, the general equation is given by

$$\frac{d^2v}{dx^2} = -\frac{1}{EI}\left[90x - \frac{20x^2}{2} + \frac{20\{x - 4\}^2}{2} - 120\{x - 6\}\right]$$

Integrate once:

$$\frac{dv}{dx} = -\frac{1}{EI}\left[\frac{90x^2}{2} - \frac{20x^3}{6} + \frac{20\{x - 4\}^3}{6} - \frac{120\{x - 6\}^2}{2} + A\right] \tag{1}$$

Integrate again:

$$v = -\frac{1}{EI}\left[\frac{90x^3}{6} - \frac{20x^4}{24} + \frac{20\{x - 4\}^4}{24} - \frac{120\{x - 6\}^3}{6} + Ax + B\right] \tag{2}$$

262

Boundary conditions $x = 0, v = 0$ give $B = 0$ from Equation (1) (no deflection at support A); $x = 8, v = 0$ give $A = -540$ from Equation (2) (no deflection at support C).

Substituting for the constants $A$ and $B$ into Equations (1) and (2) give

$$\frac{dv}{dx} = -\frac{1}{EI}\left[\frac{90x^2}{2} - \frac{20x^3}{6} + \frac{20\{x-4\}^3}{6} - \frac{120\{x-6\}^2}{2} - 540\right] \qquad (3)$$

$$v = -\frac{1}{EI}\left[\frac{90x^3}{6} - \frac{20x^4}{24} + \frac{20\{x-4\}^4}{24} - \frac{120\{x-6\}^3}{6} - 540 \times x\right] \qquad (4)$$

The slope of the beam at support A can be obtained by substituting $x = 0$ into Equation (3):

$$\frac{dv}{dx} = -\frac{1}{EI}\left[\frac{90 \times 0^2}{2} - \frac{20 \times 0^3}{6} + \frac{20\{0 \cancel{-4}\}^3}{\cancel{6}} \overset{0}{} - \frac{120\{0 \cancel{-6}\}^2}{\cancel{2}} \overset{0}{} - 540\right]$$

$$= +\frac{540}{EI}$$

**Note that in this last substitution the terms contained within the Macaulay brackets are negative and are therefore ignored.**

Similarly, the rotation at support C can be obtained by substituting $x = 8$ into Equation (3):

$$\frac{dv}{dx} = -\frac{1}{EI}\left[\frac{90 \times 8^2}{2} - \frac{20 \times 8^3}{6} + \frac{20\{8-4\}^3}{6} - \frac{120\{8-6\}^2}{2} - 540\right]$$

$$= \frac{607}{EI}$$

The deflection at B can be obtained by substituting $x = 4$ into Equation (4):

$$v = -\frac{1}{EI}\left[\frac{90 \times 4^3}{6} - \frac{20 \times 4^4}{24} + \frac{20\{4-4\}^4}{24} - \frac{120\{4 \cancel{-6}\}^3}{\cancel{6}} \overset{0}{} - 540 \times 4\right]$$

$$= \underline{1413.33/EI}$$

**Again the second term in Macaulay brackets in the above equation is neglected, as it is negative.**

## Example 11.5

A steel beam ABCD of uniform cross-section is supported and loaded as shown in Figure 11.12. The second moment of area of the cross-section of the beam is $2 \times 10^6$ mm$^4$ and Young's modulus for the steel is 200 kN/mm$^2$.

Calculate the vertical deflection at B and the slope at D.

(University of Portsmouth)

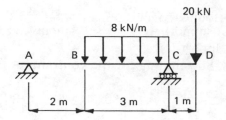

**Figure 11.12**

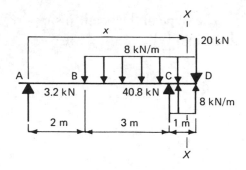

**Figure 11.13**

The values of both support reactions must be calculated and are marked on Figure 11.13. Check these reactions before proceeding. As in the previous example, the uniformly distributed load does not extend to the right-hand extremity of the beam and two equal but opposite 'dummy' loads must be introduced, as shown in Figure 11.13. This second figure is used to analyse the beam.

## Solution 11.5

The bending moment at a section $X$–$X$ close to the right-hand end of the beam is given by

$$M = 3.2x - \frac{8\{x-2\}^2}{2} + \frac{8\{x-5\}^2}{2} + 40.8\{x-5\}$$

Hence, the general equation is given by

$$\frac{\mathrm{d}^2v}{\mathrm{d}x^2} = -\frac{1}{EI}\left[3.2x - \frac{8\{x-2\}^2}{2} + \frac{8\{x-5\}^2}{2} + 40.8\{x-5\}\right]$$

Integrate once:

$$\frac{\mathrm{d}v}{\mathrm{d}x} = -\frac{1}{EI}\left[\frac{3.2x^2}{2} - \frac{8\{x-2\}^3}{6} + \frac{8\{x-5\}^3}{6} + \frac{40.8\{x-5\}^2}{2} + A\right] \quad (1)$$

Integrate again:

$$v = -\frac{1}{EI}\left[\frac{3.2x^3}{6} - \frac{8\{x-2\}^4}{24} + \frac{8\{x-5\}^4}{24} + \frac{40.8\{x-5\}^3}{6} + Ax + B\right] \quad (2)$$

264

Boundary conditions $x = 0$, $v = 0$ give $B = 0$ from Equation (2) (no deflection at support A); $x = 5$, $v = 0$ give $A = -7.93$ from Equation (2) (no deflection at support C).

Substituting for the constants $A$ and $B$ gives

$$\frac{dv}{dx} = -\frac{1}{EI}\left[\frac{3.2x^2}{2} - \frac{8\{x - 2\}^3}{6} + \frac{8\{x - 5\}^3}{6} + \frac{40.8\{x - 5\}^2}{2} - 7.93\right] \quad (3)$$

$$v = -\frac{1}{EI}\left[\frac{3.2x^3}{6} - \frac{8\{x - 2\}^4}{24} + \frac{8\{x - 5\}^4}{24} + \frac{40.8\{x - 5\}^3}{6} - 7.93 \times x\right] \quad (4)$$

The deflection at position B can be determined by substituting $x = 2$ m into Equation (4):

$$v = -\frac{1}{EI}\left[\frac{3.2 \times 2^3}{6} - \frac{8\{2 - 2\}^4}{24} + \frac{8\overset{0}{\cancel{\{2 - 5\}^2}}}{24} + \frac{40.8\overset{0}{\cancel{\{2 - 5\}^3}}}{6} - 7.93 \times 2\right]$$

$$= \frac{11.59}{EI}$$

$$= \frac{11.59}{(200 \times 10^6) \times (2 \times 10^6 \times 10^{-12})}$$

$$= 28.98 \times 10^{-3} \text{ m}$$
$$= \underline{28.98 \text{ mm}}$$

The rotation at D can be obtained by substituting $x = 6$ m into Equation (3):

$$\frac{dv}{dx} = -\frac{1}{EI}\left[\frac{3.2 \times 6^2}{2} - \frac{8\{6 - 2\}^3}{6} + \frac{8\{6 - 5\}^3}{6} + \frac{40.8\{6 - 5\}^2}{2} - 7.93\right]$$

$$= \frac{13.93}{EI}$$

$$= \frac{13.93}{(200 \times 10^6) \times (2 \times 10^6 \times 10^{-12})}$$

$$= \underline{34.83 \times 10^{-3} \text{ rad}}$$

## Example 11.6

Calculate the location of the point of maximum deflection of the beam shown in Figure 11.14 and determine the value of the maximum deflection. Take $EI = 170 \times 10^3$ kN m$^2$.

(Coventry University)

**The value of the support reaction at support A should be determined and is shown in Figure 11.15. Again this is an example where the uniformly distributed load does not extend to the right-hand side of the beam and two equal but opposite 'dummy' loads must be introduced as shown.**

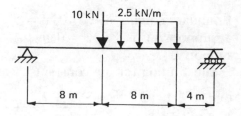

**Figure 11.14**

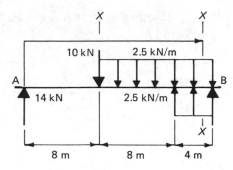

**Figure 11.15**

## Solution 11.6

The bending moment at a section $X$–$X$ close to the right-hand end of the beam is given by

$$M = 14x - 10\{x - 8\} - \frac{2.5\{x - 8\}^2}{2} + \frac{2.5\{x - 16\}^2}{2}$$

Hence, the general equation is given by

$$\frac{d^2v}{dx^2} = -\frac{1}{EI}\left[14x - 10\{x - 8\} - \frac{2.5\{x - 8\}^2}{2} + \frac{2.5\{x - 16\}^2}{2}\right]$$

Integrate once:

$$\frac{dv}{dx} = -\frac{1}{EI}\left[\frac{14x^2}{2} - \frac{10\{x - 8\}^2}{2} - \frac{2.5\{x - 8\}^3}{6} + \frac{2.5\{x - 16\}^3}{6} + A\right] \quad (1)$$

Integrate again:

$$v = -\frac{1}{EI}\left[\frac{14x^3}{6} - \frac{10\{x - 8\}^3}{6} - \frac{2.5\{x - 8\}^4}{24} + \frac{2.5\{x - 16\}^4}{24} + Ax + B\right] \quad (2)$$

Boundary conditions $x = 0$, $v = 0$ give $B = 0$ from Equation (2) (no deflection at support A); $x = 20$, $v = 0$ give $A = -682.67$ from Equation (2) (no deflection at support B).
Substituting for the constants A and B gives

$$\frac{dv}{dx} = -\frac{1}{EI}\left[\frac{14x^2}{2} - \frac{10\{x - 8\}^2}{2} - \frac{2.5\{x - 8\}^3}{6} + \frac{2.5\{x - 16\}^3}{6} - 682.67\right] \quad (3)$$

$$v = -\frac{1}{EI}\left[\frac{14x^3}{6} - \frac{10\{x-8\}^3}{6} - \frac{2.5\{x-8\}^4}{24} + \frac{2.5\{x-16\}^4}{24} - 682.67 \times x\right]$$

$$(4)$$

The position of the point of maximum deflection will occur where the slope (d$v$/d$x$) is equal to zero. This position can be determined by equating Equation (3) to zero and solving for $x$. By inspection, the point of maximum deflection will be near to the centre of the beam and to the right of the 10 kN point load. Hence, the solution will be for some value of $x$ greater than 8 m.

Equating Equation (3) to zero,

$$\frac{dv}{dx} = -\frac{1}{EI}\left[\frac{14x^2}{2} - \frac{10\{x-8\}^2}{2} - \frac{2.5\{x-8\}^3}{6} + \frac{2.5\{x-16\}^3}{6} - 682.67\right] = 0$$

As the maximum deflection will occur somewhere in the middle of the span with an approximate answer of $x = 10$ m, the first two terms in Macaulay brackets will be positive but the third term will be negative and should be ignored. Hence, the equation to be solved is given by

$$\frac{14x^2}{2} - \frac{10\{x-8\}^2}{2} - \frac{2.5\{x-8\}^3}{6} - 682.67 = 0$$

This is a cubic expression, which, by trial and error, can be shown to have the solution $x = 10.05$ m. Hence, substituting $x = 10.05$ into Equation (4),

$$v = -\frac{1}{EI}\left[\frac{14 \times 10.05^3}{6} - \frac{10\{10.05-8\}^3}{6} - \frac{2.5\{10.05-8\}^4}{24}\right.$$
$$\left. + \frac{2.5\{10.05 \cancel{-} 16\}^4}{\cancel{24}}^{\,0} - 682.67 \times 10.05\right]$$

$$\therefore v = \frac{4508}{EI}$$

$$= \frac{4508}{170 \times 10^3}$$

$$= 26.52 \times 10^{-3} \text{ m}$$
$$= \underline{26.52 \text{ mm}}$$

## Example 11.7

For the beam shown in Figure 11.16, determine the position and magnitude of the maximum deflection in terms of the flexural rigidity $EI$. The moment of 5 kNm acts about a point at 2 m from B.

(Liverpool University)

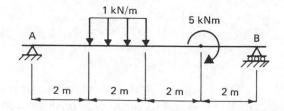

Figure 11.16

This example includes a uniformly distributed load that does not cover the whole span and, hence, two equal but opposite distributed 'dummy' loads must be added to the beam as shown in Figure 11.17. There is also a couple applied to the beam at a point 2 m from B, which makes this example different from the previous ones. The reaction at A has been calculated and shown on the diagram. Check the value of this reaction — do not forget, when taking moments about B, to include the value of the couple, which will have a moment about B although its point of application is at 2 m from B.

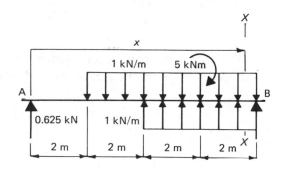

Figure 11.17

## Solution 11.7

The bending moment at a section $X$–$X$ close to the right-hand end of the beam is given by

$$M = 0.625x - \frac{1\{x - 2\}^2}{2} + \frac{1\{x - 4\}^2}{2} + 5\{x - 6\}^0$$

Note the final set of Macaulay brackets, which is associated with the applied couple. The terms in these brackets are raised to the power of zero, so that for all values of $x$ greater than 6 m this term will be positive and always equals 5 kNm.

Hence, the general equation is given by

$$\frac{d^2v}{dx^2} = -\frac{1}{EI}\left[0.625x - \frac{1\{x - 2\}^2}{2} + \frac{1\{x - 4\}^2}{2} + 5\{x - 6\}^0\right]$$

Integrate once:

$$\frac{dv}{dx} = -\frac{1}{EI}\left[\frac{0.625x^2}{2} - \frac{1\{x-2\}^3}{6} + \frac{1\{x-4\}^3}{6} + 5\{x-6\}^1 + A\right] \qquad (1)$$

Integrate again:

$$v = -\frac{1}{EI}\left[\frac{0.625x^3}{6} - \frac{1\{x-2\}^4}{24} + \frac{1\{x-4\}^4}{24} + \frac{5\{x-6\}^2}{2} + Ax + B\right] \qquad (2)$$

Boundary conditions $x = 0$, $v = 0$ give $B = 0$ from Equation (2) (no deflection at support A); $x = 8$, $y = 0$ give $A = -2.50$ from Equation (2) (no deflection at support B).

Substituting the boundary conditions,

$$\frac{dv}{dx} = -\frac{1}{EI}\left[\frac{0.625x^2}{2} - \frac{1\{x-2\}^3}{6} + \frac{1\{x-4\}^3}{6} + 5\{x-6\}^1 - 2.50\right] \qquad (3)$$

$$v = -\frac{1}{EI}\left[\frac{0.625x^3}{6} - \frac{1\{x-2\}^4}{24} + \frac{1\{x-4\}^4}{24} + \frac{5\{x-6\}^2}{2} - 2.50 \times x\right] \qquad (4)$$

**The maximum deflection will occur where the slope (d$v$/d$x$) is zero and can be obtained by equating Equation (3) to zero and solving for $x$. By inspection the point of maximum deflection will occur somewhere towards the middle of the beam within the region of the distributed load. That is, for some value of $x$ greater than 2 m and less than 4 m.**

Hence, equating Equation (3) to zero,

$$\frac{dv}{dx} = -\frac{1}{EI}\left[\frac{0.625x^2}{2} - \frac{1\{x-2\}^3}{6} + \frac{1\{x-\overset{0}{\cancel{4}}\}^3}{6} + 5\{x-\overset{0}{\cancel{6}}\}^1 - 2.50\right] = 0$$

**If it is assumed that the solution is for a value of $x$ less than 4 m, then the second and third sets of Macaulay brackets will be negative and should be ignored.**

Hence, the equation to be solved is given by

$$\frac{0.625\, x^2}{2} - \frac{1\{x-2\}^3}{6} - 2.50 = 0$$

**This is a cubic expression which, by trial and error, can be shown to have the solution $x = 2.90$ m.**

Hence, substituting $x = 2.90$ into Equation (4),

$$v = -\frac{1}{EI}\left[\frac{0.625 \times 2.90^3}{6} - \frac{1\{2.90-2\}^4}{24} + \frac{1\{2.90-\overset{0}{\cancel{4}}\}^4}{\cancel{24}} + \frac{5\{2.90-\overset{0}{\cancel{6}}\}^2}{\cancel{2}}\right.$$

$$\left. - 2.50 \times 2.90\right]$$

$$\doteq 4.74/EI$$

To determine the position of maximum deflection, it was necessary to identify the location along the span where the slope is zero. This was done in the example by inspection of the problem. However, if it is not immediately clear whether the correct region of the span has been identified, then Equation (3) could be used to calculate the slope at $x = 2$ m and $x = 4$ m. If the *sign* of the slope changes between these two positions, then it is apparent that the slope must be zero for some value of $x$ greater than 2 m and less than 4 m and, hence, the position of maximum deflection lies within this region.

**Example 11.8**

(a) A uniform cantilever of length $L$ and flexural rigidity $EI$ is subjected to a force $P$ and a hogging couple $M$ at its free end, as shown in Figure 11.18(a). By integrating the beam equation show that the deflection ($\delta$) and rotation ($\theta$) at the free end are given by

$$\delta = \frac{Pa^3}{3EI} + \frac{Ma^2}{2EI}$$

$$\theta = \frac{Pa^2}{2EI} + \frac{Ma}{EI}$$

(b) The non-uniform cantilever ABC shown in Figure 11.18(b) carries a force $P$ at its tip. AB is of length $2L$ and flexural rigidity $2EI$ and BC is of length $L$ and flexural rigidity $EI$. Using the results of part (a), show that the deflection of C is $14PL^3/3EI$.

(Birmingham University)

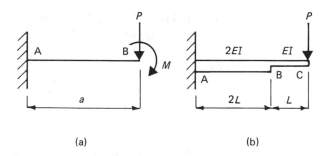

(a)                    (b)

**Figure 11.18**

*Solution 11.8*

(a) The reactions at the built-in end are shown in Figure 11.19. The bending moment at a section $X$–$X$ close to the right-hand end of the cantilever is given by

$$M = Px - (M + Pa)$$

270

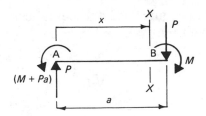

**Figure 11.19**

Hence, the general equation is given by

$$\frac{d^2v}{dx^2} = -\frac{1}{EI}[Px - (M + Pa)]$$

Integrate once:

$$\frac{dv}{dx} = -\frac{1}{EI}\left[\frac{Px^2}{2} - (M + Pa)x + A\right] \tag{1}$$

Integrate again:

$$v = -\frac{1}{EI}\left[\frac{Px^3}{6} - (M + Pa)\frac{x^2}{2} + Ax + B\right] \tag{2}$$

Boundary conditions $x = 0$, $v = 0$ give $B = 0$ from Equation (2) (no deflection at support A); $x = 0$, $dv/dx = 0$ give $A = 0$ from Equation (1) (no rotation at support A).

Hence, substituting for the constants $A$ and $B$,

$$\frac{dv}{dx} = -\frac{1}{EI}\left[\frac{Px^2}{2} - (M + Pa)x\right] \tag{3}$$

$$v = -\frac{1}{EI}\left[\frac{Px^3}{6} - (M + Pa)\frac{x^2}{2}\right] \tag{4}$$

the rotation and deflection at the free end is given by substituting $x = a$ into Equations (3) and (4), respectively:

$$\theta = \frac{dv}{dx} = -\frac{1}{EI}\left[\frac{Pa^2}{2} - (M + Pa)a\right] = \frac{Pa^2}{2EI} + \frac{Ma}{EI}$$

$$\delta = v = -\frac{1}{EI}\left[\frac{Pa^3}{6} - (M + Pa)\frac{a^2}{2}\right] = \frac{Pa^3}{3EI} + \frac{Ma^2}{2EI}$$

(b)
**To tackle this problem, two equal and opposite forces $P$ and two equal and opposite moments $PL$ should be applied to the beam at B. These will have no net effect on the behaviour of the beam. However, the problem can now be considered as the superposition of two separate problems, as shown in Figure 11.20 (a)–(d). The formulae developed in part (a) of the question can now be applied to both the separate structures shown.**

In Figure 11.20(c) the deflection ($\delta_1$) at the point of application of the load is given by:

$$\delta_1 = \frac{Pa^3}{3EI} + \frac{Ma^2}{2EI} = \frac{P(2L)^3}{3E(2I)} + \frac{(P \times L) \times (2L)^2}{2E(2I)} = \frac{7PL^3}{3EI}$$

In Figure 11.20(c) the rotation ($\theta$) at the point of application of the load is given by:

$$\theta = \frac{Pa^2}{2EI} + \frac{Ma}{EI} = \frac{P(2L)^2}{2E(2I)} + \frac{(PL) \times (2L)}{E(2I)} = \frac{2PL^2}{EI}$$

**In Figure 11.20(c) the length BC of the beam is not subjected to any bending moments and deflects as a rigid body. Hence, the deflection $\delta_2$ (for small deflections and small values of $\theta$) is given by**

$$\delta_2 = \theta \times L = \frac{2PL^2}{EI} \times L = \frac{2PL^3}{EI}$$

**In Figure 11.20(d) the moment ($P \times L$) applied at B is a 'fixing moment' which prevents B from rotating. Likewise the force ($P$) applied at B prevents B from deflecting. Thus, the length BC can be considered equivalent to a cantilevered beam section and the deflection, $\delta_3$, can be calculated from the formula developed in part (a) of the question with the applied moment $M$ equated to zero.**

Hence,

$$\delta_3 = \frac{PL^3}{3EI} + \frac{ML}{2EI} = \frac{PL^3}{3EI}$$

Hence, the total deflection of C is the sum of the three separate components of deflection:

$$\delta_C = \delta_1 + \delta_2 + \delta_3$$

$$= \frac{7PL^3}{3EI} + \frac{2PL^3}{EI} + \frac{PL^3}{3EI}$$

$$= \frac{PL^3}{EI} \left( \frac{7}{3} + 2 + \frac{1}{3} \right)$$

$$= \underline{14PL^3/3EI}$$

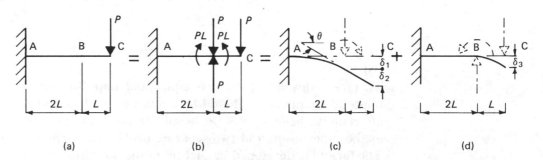

(a)          (b)          (c)          (d)

**Figure 11.20**

272

# 11.5    Problems

**11.1**    A simply supported beam AB is 6 m long and carries two point loads $P$ at its quarter-span points, as shown in Figure P11.1. A support C at the mid-span point is 12 mm below the beam before the loads are applied. Calculate the magnitude of the loads $P$ so that the beam just touches the support at C. Take $E = 200$ kN/mm$^2$ and $I = 165 \times 10^6$ mm$^4$.

(Manchester University)

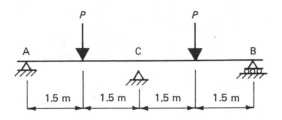

**Figure P11.1**

**11.2**    Find the deflection under the point load and at the mid-point of AB of the beam shown in Figure P11.2 in terms of the flexural rigidity $EI$.

(University of Westminster)

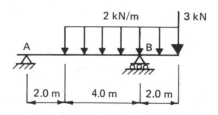

**Figure P11.2**

**11.3**    The cantilever ABC shown in Figure P11.3 is of uniform cross-section and is subjected to a uniformly distributed load over one-half of the span. Determine the deflection of the cantilever at C. The flexural rigidity of the beam is $EI$.

(Salford University)

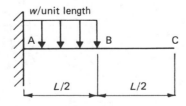

**Figure P11.3**

**11.4** Figure P11.4 shows the loading on a beam simply supported over a span of 4 m. If the flexural rigidity $EI$ is constant along the length of the beam and is equal to 2000 kNm$^2$, determine:

(i)   the deflection at end A;
(ii)  the deflection at the mid-span at C;
(iii) the slope at the support B.

(Nottingham Trent University)

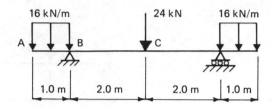

**Figure P11.4**

**11.5** Figure P11.5 shows a beam ABCD which is of uniform cross-section throughout. It is supported on a knife edge at A and a knife edge and roller system at C. It carries point loads of 60 kN and 50 kN at B and D, respectively.

The maximum permitted vertical deflection due to the simultaneous application of both point loads is 20 mm. Determine the minimum value of flexural rigidity $EI$ for the section to comply with this criterion.

(Sheffield University)

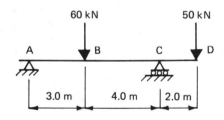

**Figure P11.5**

**11.6** The loading on a simply supported beam of uniform cross-section is shown in Figure P11.6. If Young's modulus is 200 kN/mm$^2$ and the second moment of area is $75 \times 10^6$ mm$^4$, calculate the central deflection.

(University of Hertfordshire)

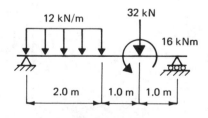

**Figure P11.6**

## 11.6 Answers to Problems

**11.1** 64 kN

**11.2** $23.55/EI$, $2.92/EI$

**11.3** $7wL^4/384EI$

**11.4** (i) $-3$ mm; (ii) 8 mm; (iii) $4 \times 10^{-3}$ rad

**11.5** $12.86 \times 10^3$ kNm$^2$ (maximum span deflection between A and B = $128.68/EI$; maximum cantilever deflection = $257.15/EI$; cantilever critical)

**11.6** 4.09 mm

# 12 Strain Energy and Virtual Work

## 12.1 Contents

Calculation of the deflection of the joints of statically determinate pin-jointed frames using the methods of virtual work and strain energy ● Strain energy methods for the solution of problems involving springs.

Many first-year courses in structural mechanics will introduce the strain energy and virtual work methods and will apply these techniques to problems involving the deflection of simple pin-jointed frames and the calculation of the deformation and forces within elastic springs. Both of these methods form the basis of a number of powerful analytical techniques and will usually subsequently be studied at a more advanced level.

In this chapter a number of typical first-year level examination questions are given, based on one or the other of these two methods.

## 12.2 The Fact Sheet

**(a) Strain Energy Stored in an Elastic Bar**

The strain energy ($U$) stored in an elastic bar of cross-sectional area $A$, length $L$ and elastic modulus $E$ when subjected to an axial load $P$ is given by

$$U = \frac{P^2 L}{2AE} \qquad (1)$$

**(b) Pin-jointed Plane Frames: Deflection Calculation Using Strain Energy**

By equating the work done by the external loading to the gain in internal strain energy stored within a pin-jointed frame, the following expression is obtained:

$$\sum \frac{W \times \delta}{2} = \sum \frac{P^2 L}{2AE} \qquad (2)$$

where the expression on the right-hand side represents the total strain energy stored within all the members of the frame. The expression on the left-hand side is the total work done by the external loading. $W$ is a load applied at a joint of the frame and $\delta$ is the corresponding deflection of the joint in the direction of the load.

Equation 2 can be used to calculate joint deflections but is usually limited to problems where there is only one unknown deflection at the point of application of a single load.

### (c)   Strain Energy Stored within an Elastic Spring

The force $P$ in an elastic spring of *stiffness* $K$ is given by:

$$P = K \times e$$

where $e$ is the shortening or extension of the spring.

The *spring stiffness* ($K$) is equivalent to the *axial stiffness* ($EA/L$) of an elastic bar. The strain energy stored within a spring is given by

$$U = \tfrac{1}{2}P \times e = \tfrac{1}{2}P \times \frac{P}{K} = \frac{P^2}{2K} \tag{3}$$

where $P$ is the force within the spring.

### (d)   Pin-jointed Plane Frames: Deflection Calculation Using Virtual Work

The principle of virtual work states that *if a system of forces is in equilibrium and undergoes any arbitrary virtual displacement, then the total work done by the system of forces is zero.*

The application of this principle to pin-jointed frames gives the general equation:

$$\delta = \sum^{n} P_{II}e = \sum^{n} P_{II}\frac{(P_{I}L)}{AE} \tag{4}$$

where
   $e$ = the change in member length due to the actual load system;
   $\delta$ = the deflection of a joint in the loaded frame;
   $P_I$ = the member axial forces due to the actual load system (force set I);
   $P_{II}$ = the member axial forces due to a unit load applied at the joint where the displacement $\delta$ is required and in the direction of the required deflection (force set II).

This equation is much more powerful than the corresponding strain energy equation 2 and is more readily applicable to a wider range of problems, as will be demonstrated in the worked examples.

## 12.3  Symbols, Units and Sign Conventions

$\delta$ = deflection (mm)
$e$ = change in member length (mm)
$A$ = area of member cross-section (mm$^2$)
$E$ = Young's modulus of elasticity (kN/mm$^2$)
$K$ = spring stiffness (kN/m)
$L$ = length (m)
$P$ = member axial force (kN)
$U$ = strain energy (J)

Tensile forces are taken as positive.
Axial extensions are taken as positive.
Deflections are positive when they are in the same direction as the corresponding force.

## 12.4  Worked Examples

### Example 12.1

A rigid beam AB is shown in Figure 12.1. The beam is simply supported at A and by a spring of stiffness 200 kN/m at B. A mass of 50 kg is dropped from a height of 1.5 m onto the beam at C, which is 3 m from A. Assuming that the beam has negligible mass and that there is no loss of energy calculate:

(i)   the maximum deflection at B;
(ii)  the force induced in the spring.

(University of Portsmouth)

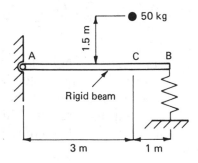

**Figure 12.1**

Figure 12.2 shows the deflected beam and the deformed spring after the body comes instantaneously to rest after impacting with the beam. As the beam is stated to be rigid it remains straight and hence if the deflection at B is $\delta$ then by simple proportion the deflection at C is $\frac{3}{4}\delta$.

278

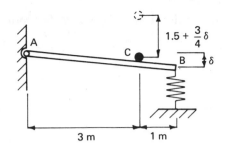

**Figure 12.2**

### Solution 12.1

Potential energy lost by mass = mg× height through which it falls
$$= 50 \times 9.81 \times (1.5 + \tfrac{3}{4}\delta)$$
$$= 490.50 \ (1.5 + \tfrac{3}{4}\delta) \ \text{J}$$

**Note that the energy is expressed in joules (newton metres). Hence, $\delta$ in this equation is in metres.**

$$\text{Strain energy in spring} = \frac{P^2}{2K} = \frac{P^2}{2 \times 200 \times 10^3} = 2.5 \times P^2 \times 10^{-6} \ \text{J}$$

**Again the energy is expressed in joules and, hence, $P$ in this equation will be expressed in newtons. The stiffness, $K$, has been converted to N/m.**

Equating the potential energy lost by the mass to the strain energy gained by the spring,

$$490.50 \ (1.5 + \tfrac{3}{4}\delta) = 2.5 \times P^2 \times 10^{-6} \tag{1}$$

But the force in the spring and the deflection of the spring are related by the spring stiffness—that is,

$$P = K \times \delta$$

or

$$\delta = P/K = P/200\ 000 = 5P \times 10^{-6}$$

Hence, substituting into Equation (1),

$$490.50 \ (1.5 + \tfrac{3}{4} \times 5 \times P \times 10^{-6}) = 2.5 \times P^2 \times 10^{-6}$$

which simplifies to give

$$P^2 - 735.75P - 294.3 \times 10^6 = 0$$

which has the solution

$$P = 17526.99 \ \text{N}$$
$$= \underline{17.53 \ \text{kN}}$$

The deflection at B is the deflection of the spring. Hence, as $P = K \times \delta$, then

$$\delta = \frac{P}{K}$$

$$= \frac{17\,526.99}{200 \times 10^3}$$

$$= 87.63 \times 10^{-3} \text{ m}$$

$$= \underline{87.63 \text{ mm}}$$

**Example 12.2**

The pin-jointed framework shown in Figure 12.3 is connected by pins to a wall at A and F and carries vertical loads of 36 kN and 60 kN at C and E, respectively.

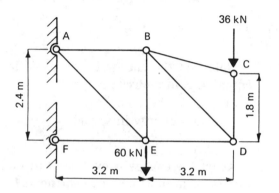

**Figure 12.3**

(i) Calculate the total strain energy stored within the frame.
(ii) If the vertical deflection at C is 15.14 mm, what would be the vertical deflection at E?

Young's Modulus ($E$) for all members = 200 kN/mm$^2$. The cross-sectional area of AE and EF is 1800 mm$^2$ and for all other members the cross-sectional area is 600 mm$^2$.

(Coventry University)

   **To calculate the strain energy, it is first necessary to determine the forces within all members, as shown in Figure 12.4. The reader should check these values, using the method of resolution at joints.**

*Solution 12.2*

The strain energy is given by the formula $U = \Sigma P^2 L / 2AE$. The calculations are best set out in tabular form:

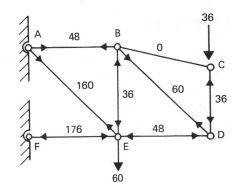

**Figure 12.4**

| Member | $P$ (kN) | Length (m) | Area (mm$^2$) | $P^2L/A$ (N$^2$/m $\times 10^{12}$) |
|---|---|---|---|---|
| AB | 48 | 3.2 | 600 | 12.288 |
| BC | 0 | 3.3 | 600 | 0 |
| CD | −36 | 1.8 | 600 | 3.888 |
| DE | −48 | 3.2 | 600 | 12.288 |
| EF | −176 | 3.2 | 1800 | 55.068 |
| AE | 160 | 4.0 | 1800 | 56.889 |
| BE | −36 | 2.4 | 600 | 5.184 |
| BD | 60 | 4.0 | 600 | 24.000 |

$$\sum \frac{P^2L}{A} = 169.606 \times 10^{12}$$

Hence,

$$U = \sum \frac{P^2L}{2AE} = \frac{1}{2E} \sum \frac{P^2L}{A} = \frac{169.606 \times 10^{12}}{2 \times 200 \times 10^9}$$

$$= 424.02 \text{ J}$$

**Note carefully the units in these calculations. All units have been converted to newtons and metres, so that the strain energy is calculated in joules (newton metres).**

The external work done by the two loads is given by

$$\text{work done} = \tfrac{1}{2}\,(W_C\delta_C + W_E\delta_E)$$
$$= \tfrac{1}{2}\,(36\,000 \times 15.14 \times 10^{-3} + 60\,000 \times \delta_E \times 10^{-3}) \text{ J}$$

Hence, equating work done by the external loads to the strain energy stored,

$$\tfrac{1}{2}\,(36\,000 \times 15.14 \times 10^{-3} + 60\,000 \times \delta_E \times 10^{-3}) = 424.02$$

Hence,

$$\delta_E = \underline{5.05 \text{ mm}}$$

## Example 12.3

All members of the plane truss shown in Figure 12.5 have cross-sectional area $A$ and Young's Modulus $E$. The length of all horizontal and vertical members is $L$.

The truss is simply supported over a span of $2L$ as shown. Find an expression for the vertical displacement of C when a horizontal load $F$ is applied at D in the plane of the truss.

(Aberdeen University)

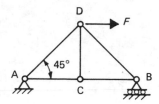

**Figure 12.5**

## Solution 12.3

**This is an example which is best solved by virtual work methods rather than strain energy techniques, as the required deflection is not in the direction of the given single applied force. However, it is still necessary to calculate the member forces (Set I), as shown in Figure 12.6(a). This is best done by the method of resolution at joints. Figure 12.6(b) shows the member forces (Set II) due to a single unit load applied at joint C in the direction of the required vertical deflection. Again the calculations are best presented in tabular form. The fourth column of the table gives the change of member lengths due to the actual load system calculated from the expression $e = PL/AE$.**

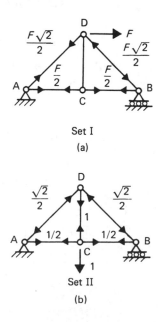

**Figure 12.6**

| Member | Length | $P$ (Set I) | $e$ (PL/AE) (Set I) ($\times FL/AE$) | $P_{II}$ (Set II) | $P_{II} \times e$ (Set II) $\times$ (Set I) ($\times FL/AE$) |
|--------|--------|------------|-----------|----------|-----------|
| AD | 1.414L | 0.707F | 1.0 | −0.707 | −0.707 |
| DB | 1.414L | −0.707F | −1.0 | −0.707 | 0.707 |
| BC | 1.000L | 0.500F | 0.5 | 0.500 | 0.250 |
| AC | 1.000L | 0.500F | 0.5 | 0.500 | 0.250 |
| CD | 1.000L | 0 | 0 | 1.000 | 0 |

$$\sum^{n} P_{II}e = 0.500 \times \frac{FL}{AE}$$

Hence,

$$\text{deflection of C} = \delta_C = \sum^{n} P_{II}e = 0.5\frac{FL}{AE}$$

## Example 12.4

The members in the pin-jointed framework shown in Figure 12.7 all have a cross-sectional area of 1000 mm$^2$ and a Modulus of Elasticity of 200 kN/mm$^2$. Find the vertical deflection of joint C when vertical loads are applied at joints A and E as shown.

(Salford University)

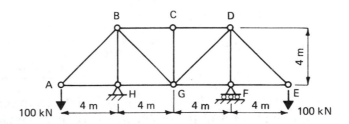

**Figure 12.7**

## Solution 12.4

**This is another problem which is easily solved by the use of virtual work. However, note the symmetry of the problem, which reduces the amount of necessary calculation. The solution is set out as in the previous problem. Figures 12.8(a) and 12.8(b) show the actual member forces and those due to a unit vertical load applied at joint C. Again the calculations are presented in tabular form:**

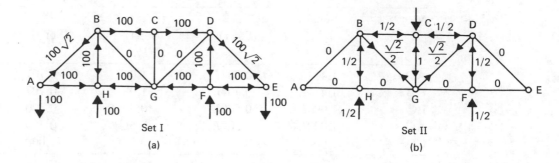

**Figure 12.8**

| Member | Length (m) | P (Set I) (kN) | e (PL/AE) (Set I) (mm) | $P_{II}$ (Set II) (kN) | $P_{II} \times e$ (Set II) × (Set I) (kNmm) |
|---|---|---|---|---|---|
| | | | $\dfrac{141.42 \times 5.657 \times 10^3}{1000 \times 200}$ | | |
| AB | 5.657 | 141.42 | = 4.0 | 0 | 0 |
| BC | 4.000 | 100.00 | 2.0 | −0.500 | −1.0 |
| CG | 4.000 | 0 | 0 | −1.000 | 0 |
| GH | 4.000 | −100.00 | −2.0 | 0 | 0 |
| AH | 4.000 | −100.00 | −2.0 | 0 | 0 |
| BH | 4.000 | −100.00 | −2.0 | −0.500 | 1.0 |
| BG | 5.657 | 0 | 0 | 0.707 | 0 |

Hence,

$$\text{deflection of C} = \delta_C = \sum^{n} P_{II}e = 2 \times (0 - 1.0 + 0 + 0 + 0 + 1.0 + 0) + 0$$

$$= \underline{0.0 \text{ mm}}$$

Note that in the table only half the frame has been considered, to account for the symmetry of the problem. Hence, in the final equation all terms corresponding to all members except CG have been doubled to account for the right-hand half of the frame.

Rather a surprising result! In fact, the cross-sectional area and modulus of elasticity were not really needed, as, provided that all the members have the same properties, the answer would have been the same whatever the values of these two properties.

Inspection of the problem would have shown that many of the members have zero force in them under the action of either the actual load system or the unit load system. Hence, the calculations could easily have been shortened to take this into account.

*Units* are important in this problem. In the table the units of force have been kept in kN and units of length have been converted to mm, so that the final answer can be quoted in mm.

## Example 12.5

A plane pin-jointed frame is shown in Figure 12.9. All members have the same Modulus of Elasticity $E$ and cross-sectional area $A$. Using the principle of Virtual Work calculate both the vertical displacement and horizontal displacement of point G when a 10 kN vertical load is applied at G.

(Sheffield University)

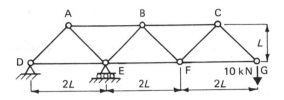

**Figure 12.9**

## Solution 12.5

**The calculations for this problem are set out in a similar way to the previous two problems. However, as two deflections are to be calculated, the table of calculations contains two extra columns as shown. Refer to Figure 12.10.**

| Member | Length | $P$ (Set I) | $e$ $(PL/AE)$ (Set I) $(\times L/AE)$ | $P_{II}$ (Set II) | $P_{II} \times e$ $(\times L/AE)$ | $P_{II}$ (Set II) | $P_{II} \times e$ $(\times L/AE)$ |
|---|---|---|---|---|---|---|---|
| | | | | Horizontal deflection | | Vertical deflection | |
| AB | 2.000$L$ | 40.000 | 80 | 0 | 0 | 4.000 | 320.000 |
| BC | 2.000$L$ | 20.000 | 40 | 0 | 0 | 2.000 | 80.000 |
| CG | 1.414$L$ | 14.142 | 20 | 0 | 0 | 1.414 | 28.284 |
| FG | 2.000$L$ | −10.000 | −20 | 1 | −20 | −1.000 | 20.000 |
| EF | 2.000$L$ | −30.000 | −60 | 1 | −60 | −3.000 | 180.000 |
| DE | 2.000$L$ | −20.000 | −40 | 1 | −40 | −2.000 | 80.000 |
| AD | 1.414$L$ | 28.284 | 40 | 0 | 0 | 2.828 | 113.137 |
| AE | 1.414$L$ | −28.284 | −40 | 0 | 0 | −2.828 | 113.137 |
| BE | 1.414$L$ | −14.142 | −20 | 0 | 0 | −1.414 | 28.284 |
| BF | 1.414$L$ | 14.142 | 20 | 0 | 0 | 1.414 | 28.284 |
| FC | 1.414$L$ | −14.142 | −20 | 0 | 0 | −1.414 | 28.284 |
| | | | | | −120 | | 1019.126 |

Hence,

$$\text{horizontal deflection of G} = \delta_{GH} = \sum^{n} P_{II}e = \frac{-120L}{AE} \text{ (i.e. to the left)}$$

$$\text{vertical deflection of G} = \delta_{GV} = \sum^{n} P_{II}e = \frac{1019.13L}{AE}$$

285

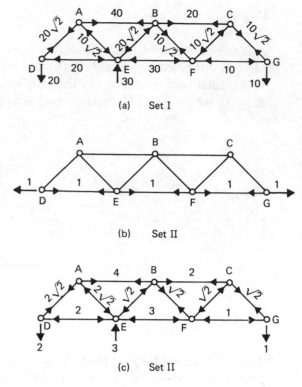

(a)  Set I

(b)  Set II

(c)  Set II

**Figure 12.10**

## Example 12.6

(a)  Determine the force in each of the six diagonals of the framework shown in Figure 12.11 under the loading indicated. The diagonals are of length $\sqrt{2}a$ and all other members are of length $a$.

(b)  The diagonals have elastic stiffness under tension/compression of 20 kN/mm and all other members are completely rigid under this sort of loading. The members do not buckle. By use of virtual work determine the horizontal component of displacement of the point of application of the topmost force.

(Cambridge University)

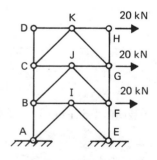

**Figure 12.11**

*Solution 12.6*

(a)
**Part (b) of the question states that the non-diagonal members are completely rigid and, hence, will not change in length under the given load system. These members will therefore not contribute to the virtual work equation and the forces in these members are not required. Part (a) of the question is giving a 'hint' to the solution of part (b), as only the forces in the diagonals are asked for. To calculate the forces in the diagonal members, it is probably easiest to use the method of sections in combination with some resolution at joints rather than to determine all the member forces by resolving at all the joints.**

Hence, for the top two diagonal members, resolving vertically at joint K, ($\Sigma V = 0$)

$$-P_{CK} \sin45° - P_{GK} \sin45° = 0$$
$$P_{CK} = -P_{GK}$$

Resolving horizontally for the section (see Figure 12.12)

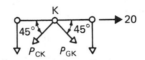

**Figure 12.12**

($\Sigma H = 0$)
$$-P_{CK} \cos45° + P_{GK} \cos45° + 20 = 0$$

Solving,

$$P_{CK} = 14.142 \text{ kN}$$
$$P_{GK} = -14.142 \text{ kN}$$

For the middle two diagonal members: resolving vertically at joint J, ($\Sigma V = 0$)

$$-P_{BJ} \sin45° - P_{FJ} \sin45° = 0$$
$$P_{BJ} = -P_{FJ}$$

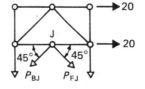

**Figure 12.13**

Resolving horizontally for the section (see Figure 12.13), ($\Sigma H = 0$)

$$-P_{BJ} \cos45° + P_{FJ} \cos45° + 40 = 0$$

Solving,

$$P_{BJ} = 28.284 \text{ kN}$$
$$P_{FJ} = -28.284 \text{ kN}$$

And similarly,

$$P_{AI} = 42.426 \text{ kN}$$
$$P_{EI} = -42.426 \text{ kN}$$

(b)

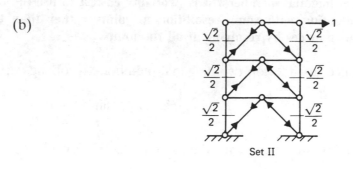

Set II

Figure 12.14

**Figure 12.14 shows the member forces in the diagonals due to a unit horizontal load applied at joint H, where the horizontal deflection is required. Again these forces are best calculated using the method of sections. The calculations for deflection are set out below in tabular form. As explained in the comment to part (a) of this question, only the diagonal members are included in the calculations, as the other members are completely rigid and do not deform. In this question the member stiffness is given as 20 kN/mm and, hence, the extension of a member ($e$ in mm) can be calculated by dividing the member force ($P$ in kN) by the stiffness (20 kN/mm).**

| Member | $P$ (Set I) (kN) | $e$ ($P/20$) (Set I) (mm) | $P_{II}$ (Set II) (kN) | $P_{II}e$ (Set II) × (Set I) |
|--------|------|------|------|------|
| CK | 14.142 | 0.707 | 0.707 | 0.50 |
| GK | −14.142 | −0.707 | −0.707 | 0.50 |
| BJ | 28.284 | 1.414 | 0.707 | 1.00 |
| FJ | −28.284 | −1.414 | −0.707 | 1.00 |
| AI | 42.426 | 2.121 | 0.707 | 1.50 |
| EI | −42.426 | −2.121 | −0.707 | 1.50 |

$$\sum^{n} P_{II}e = 6.00$$

Hence,

horizontal deflection at G $= \delta_G = \sum^{n} P_{II}e = \underline{6.00 \text{ mm}}$

## Example 12.7

The pin-jointed three-dimensional frame shown in Figure 12.15 is attached to a vertical wall at points B, C and D. The three members have the same

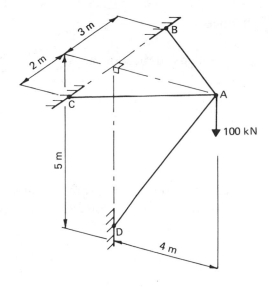

**Figure 12.15**

cross-sectional area and the modulus of elasticity of the material is 200 kN/mm².

If the vertical deflection at the apex A is not to exceed 3 mm due to the application of a vertical force of 100 kN at A, determine the minimum cross-sectional area that may be used for the members.

(Nottingham Trent University)

*Solution 12.7*

**This is a three-dimensional problem where the member forces have to be determined by some appropriate method such as the method of tension coefficients. An example using the method of tension coefficients is given in Chapter 2 and the reader should check that the answers tabulated below for the member forces in this problem are correct. Once the member forces are calculated, the rest of the calculations can be set out in the usual way. As the vertical deflection of A is required, the member forces due to a unit load applied at A must be calculated. However, as the member forces due to a load of 100 kN have already been calculated, the member forces due to a unit load (Set II) are simply determined by dividing the Set I forces by 100.**

| Member | Length (m) | $P$ (Set I) (kN) | $e$ $(PL/AE)$ (Set I) $(\times 1/AE)$ | $P_{II}$ (Set II) (kN) | $P_{II}e$ (Set II) × (Set I) $(\times 1/AE)$ |
|--------|-----------|------------------|---------------------------------------|------------------------|----------------------------------------------|
| AB | 5.00 | 40.00 | 200.00 | 0.40 | 80.00 |
| AC | 4.47 | 53.66 | 239.86 | 0.54 | 129.52 |
| AD | 6.40 | −128.06 | −819.58 | −1.28 | 1049.06 |

$$\sum^{II} P_{II}e = 1258.58$$

Hence,

$$\text{vertical deflection at A} = \delta_A = \sum^n P_{II}e = \frac{258.58}{AE} = \frac{1258.58}{A \times 200 \times 10^6}$$

$$= \frac{6.29}{A \times 10^6} \text{ m}$$

$$= \frac{6.29}{A \times 10^3} \text{ mm}$$

where the area $(A)$ is expressed in square metres.

But the maximum deflection of joint A is limited to 3 mm. Hence,

$$\frac{6.29}{A \times 10^3} < 3$$

or

$$A > \frac{6.29}{3 \times 10^3}$$

$$> 2.097 \times 10^{-3} \text{ m}^2$$
$$> 2.097 \times 10^{-3} \times 10^6 \text{ mm}^2$$
$$> \underline{2097 \text{ mm}^2}$$

**It is worth noting that, as this structure is subjected to a single point load and the deflection at the point of application of the load is to be calculated, the problem could have been solved using the strain energy method. The reader should repeat the problem using the strain energy approach.**

## 12.5  Problems

12.1 (a) Prove that if a load $W$ is suddenly applied to a structural member, the effect is the same as if a load $2W$ is gradually applied to that member.
   (b) During a concreting operation a skip opens prematurely, allowing 2500 kg of concrete to fall through a distance of 3 m onto a timber panel supported by four props.

   Assuming that each prop carries a quarter of the load and that the stiffness of the props is $27 \times 10^6$ N/m calculate:

   (i)   the instantaneous compression in each prop;
   (ii)  the instantaneous force in each prop.

   (Coventry University)

12.2 Figure P12.1 shows a plane truss for temporary bracing of a structure during construction. If the cross-section of all members is 1000 mm² and Young's Modulus is 200 kN/mm², find the horizontal displacement of the joint A under the loading shown.

   (Sheffield University)

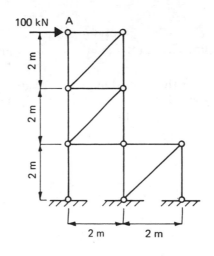

**Figure P12.1**

**12.3** Figure P12.2 shows a pin-jointed frame in which the cross-sectional area of all vertical and horizontal members is $A$ and that of all diagonal members is $A\sqrt{2}$ and all have the same Modulus of Elasticity, $E$. Determine, in terms of $W$, $A$, $L$ and $E$ the horizontal deflection of D due to the application of the vertical load $W$ and calculate its numerical value when $W = 50$ kN, $A = 11$ cm$^2$, $L = 3$ m and $E = 200$ kN/mm$^2$.

(Nottingham University)

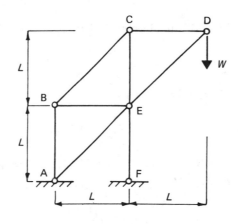

**Figure P12.2**

**12.4** A truss is loaded as shown in Figure P12.3. Calculate the horizontal and vertical deflection of joint A if the cross-sectional areas of the compression members is 1000 mm$^2$ and that of the tension members is 750 mm$^2$. Take $E = 200$ kN/mm$^2$

(Coventry University)

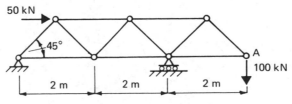

**Figure P12.3**

**12.5** Figure P12.4 shows a pin-jointed arch truss supporting a vertical load $W$ at the point C. Determine the horizontal and vertical components of the reactions at A and D.

Another loading case results in the members AB, BC, and CD each shortening by a small length, $e$. Assuming that all other members are effectively unchanged in length, find the resulting vertical displacement of point C. What is the corresponding horizontal displacement of C?

(Cambridge University)

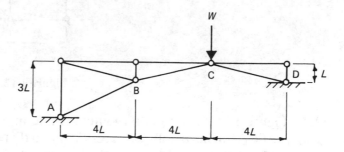

**Figure P12.4**

# 12.6  Answers to Problems

**12.1**  (b) 18.68 mm, 504.48 kN

**12.2**  27.48 mm

**12.3**  $10WL/AE$, 6.82 mm

**12.4**  $\delta_H = -2.50$ mm, $\delta_V = 14.82$ mm

**12.5**  $V_A = 0.6W$, $H_A = 1.6W$, $V_D = 0.4W$, $H_D = -1.6W$, $5.91e$ (downwards), $0.45e$ (to the left)

# Index